管理會計（雙語教材）

主　編 ◎ 張紅云、韓衛華
副主編 ◎ 周承彥、朱　丹、陳特特

前　言

　　管理會計是管理學中會計學科下的一門邊緣學科，主要研究如何取得有效的信息，為企業管理者進行計劃、控制及決策服務。隨著中國市場經濟的不斷發展，市場競爭日益激烈，面對不斷變化的市場環境，企業的風險與機會並存。因此，企業必須改善經營模式，借助先進的管理會計技術和工具，提高經營管理水平，實現企業價值的持續增長，使企業在經濟發展的新時期和全球化趨勢中立於不敗之地。伴隨著工業4.0時代的到來，實體經濟與信息系統的高度融合使智能化生產、物流、服務融為一體，使經濟信息的統一採集、精確計算、遠程操控、實時分析成為可能，管理會計的理論研究與實踐得到了飛躍式發展，管理會計4.0時代悄然而至。在此背景下，財政部提出大力發展管理會計，全面推進管理會計體系建設，為社會經濟發展提供有效的服務和保障。而管理會計的理論與技術方法的學習及傳播對企業經營管理人員提高管理效率具有特別現實的意義。隨著經濟的全球化，越來越多的企業開展對外貿易、對外投資，參與到國際競爭中去，經營管理人員用國際上通用的語言來分析問題、開展工作、進行交流的能力也日益受到重視。順應這一趨勢變化，借助雙語教材展開雙語教學是各層次管理人員獲取相應能力的有效途徑。

　　管理會計是一門應用性很強的學科，本書在闡述基本理論知識的基礎上，將重點、難點與實例相結合，深入講解，力求清楚透澈，突出重要內容、核心內容。本書在編寫上有如下特點：

　　1. 內容上簡練實用、重點突出。本書對管理會計的相關理論和技術方法進行合理的選擇與整理，突出其基本的、重點的、實用的內容，這使得本書在內容上更實用化，使學習者能系統而有效地掌握管理會計的重要知識和技術方法。

　　2. 結構上中英文搭配。對重要的概念、理論和方法以及難以理解部分加註中文，予以解釋，使學習者便於理解、掌握。

　　3. 注重理論與實踐相結合。每章的重點內容和技術方法均編有例題加以解釋，說明其應用，章節末設置有練習題，有利於提高學習者理解知識、應用相關方法的能力。

本書由張紅雲、韓衛華任主編，周承彥、朱丹、陳特特任副主編。張紅雲、韓衛華負責本書的總體框架設計、各章初稿的修改和全書的總撰與定稿。全書共分10章，編寫的具體分工為：第一章、第二章由朱丹編寫，第三章、第四章由周承彥編寫，第五章、第六章、第九章由張紅雲編寫，第七章由陳特特編寫，第八章、第十章由韓衛華編寫。

本書可作為高等院校會計學、財務管理、工商管理、會計電算化等經濟管理類專業的教材，也可以作為會計實務工作人員、經濟管理人員的參考讀物。

本書在編寫過程中參考和引用了大量的國內外專家學者的著作，在此表示誠摯謝意。

由於編者水平所限，書中難免出現錯誤和紕漏，懇請各位讀者批評指正。

編　者

Contents

Chapter 1 Management Accounting Overview ... 1)
    1.1 Management Accounting, Finance Accounting and Cost Accounting ... 1)
    1.2 Management Information ... 3)
    1.3 Cost and Cost Classification ... 5)

Chapter 2 Cost Accounting Principles ... 15)
    2.1 Accounting for Material ... 15)
    2.2 Accounting for Labor ... 22)
    2.3 Accounting for Overheads ... 30)

Chapter 3 Absorption Costing and Marginal Costing ... 41)
    3.1 Absorption Cost and Absorption Costing ... 41)
    3.2 Marginal Cost and Marginal Costing ... 44)
    3.3 Absorption Costing, Marginal Costing and the Calculation of Profit ... 45)
    3.4 Reconciling Profits ... 49)

Chapter 4 Cost Accounting Methods ... 54)
    4.1 Job Costing ... 54)
    4.2 Batch Costing ... 61)
    4.3 Process Costing ... 62)
    4.4 Service Costing ... 71)

Chapter 5 Activity Based Costing and Other Cost Management Tools ... 79)
    5.1 Activity Based Costing (ABC) ... 79)
    5.2 Target Costing ... 81)
    5.3 Life Cycle Costing ... 83)
    5.4 Total Quality Management (TQM) ... 85)

Chapter 6 Cost Behavior and Cost-Volume-Profit (CVP) Analysis ... 88)
    6.1 Cost Behavior ... 88)
    6.2 Basic CVP Analysis ... 92)
    6.3 Using CVP to Break-even Analysis ... 95)

6.4	Using CVP to Plan Profit	100)
6.5	Using CVP for Sensitivity Analysis	102)
6.6	Effect of Sales Mix on CVP Analysis	105)

Chapter 7 Budgeting 113)

7.1	Budgeting Overview	113)
7.2	The Components of the Master Budgeting	117)
7.3	Preparing the Operating Budget	119)
7.4	Preparing the Financial Budget	130)
7.5	Using Information Technology for Sensitivity Analysis, Variance Analysis and Rolling Up Unit Budgets	135)

Chapter 8 Standard Costing and Variance Analysis 141)

8.1	Standard Costing Overview	141)
8.2	Establishing Cost Standard	142)
8.3	Basic Variance Analysis	146)
8.4	Sales Variance	155)
8.5	Reconciliation of Budget and Actual Profit Under Standard Absorption Costing	157)
8.6	Reconciliation of Budget and Actual Profit or Contribution Under Standard Marginal Costing	157)

Chapter 9 Time Value of Money and Capital Investment Appraisal 163)

9.1	Capital Investment and Capital Expenditure Budget	163)
9.2	Time Value of Money	165)
9.3	Using Non-discounted Cash Flow Models to Make Capital Investment Appraisal	(174)
9.4	Using Discounted Cash Flow Models to Make Capital Investment Appraisal	(178)
9.5	Comparing Capital Investment Appraisal Models	186)

Chapter 10 Performance Measurement 191)

10.1	Performance Measurement Overview	191)
10.2	Application of Performance Measurement for Organization	193)
10.3	Application of Performance Measurement for Responsibility Centers	197)
10.4	Economic Value Added	199)
10.5	Balanced Scorecard	202)

Appendix: Present Value Tables and Future Value Tables 209)

目　錄

第 1 章　管理會計概述　(1)
　1.1　管理會計、財務會計和成本會計　(1)
　1.2　管理信息　(3)
　1.3　成本和成本分類　(5)

第 2 章　成本核算原則　(15)
　2.1　材料核算　(15)
　2.2　人工核算　(22)
　2.3　間接費用核算　(30)

第 3 章　完全成本法和變動成本法　(41)
　3.1　完全成本和完全成本法　(41)
　3.2　變動成本和變動成本法　(44)
　3.3　完全成本法和變動成本法下利潤的計算　(45)
　3.4　利潤的調節　(49)

第 4 章　成本核算方法　(54)
　4.1　訂單成本法　(54)
　4.2　分批成本法　(61)
　4.3　分步成本法　(62)
　4.4　服務成本法　(71)

第 5 章　作業成本法和其他成本管理工具　(79)
　5.1　作業成本法（ABC）　(79)
　5.2　目標成本法　(81)
　5.3　生命週期成本法　(83)
　5.4　全面質量管理（TQM）　(85)

第 6 章　成本習性和本量利（CVP）分析　(88)
　6.1　成本習性　(88)
　6.2　基本的本量利分析　(92)

6.3　本量利分析的運用-盈虧平衡分析　　　　　　　　　　　　　　　（95）
　　6.4　本量利分析的運用-利潤規劃分析　　　　　　　　　　　　　　（100）
　　6.5　本量利分析的運用-敏感分析　　　　　　　　　　　　　　　　（102）
　　6.6　本量利分析中銷售組合的影響　　　　　　　　　　　　　　　　（105）

第7章　預算　　　　　　　　　　　　　　　　　　　　　　　　　　　（113）
　　7.1　預算概述　　　　　　　　　　　　　　　　　　　　　　　　　（113）
　　7.2　預算體系的構成　　　　　　　　　　　　　　　　　　　　　　（117）
　　7.3　業務預算的編制　　　　　　　　　　　　　　　　　　　　　　（119）
　　7.4　財務預算的編制　　　　　　　　　　　　　　　　　　　　　　（130）
　　7.5　使用信息技術進行靈敏度分析，差異分析及編制單位預算　　　　（135）

第8章　標準成本法和差異分析　　　　　　　　　　　　　　　　　　　（141）
　　8.1　標準成本法概述　　　　　　　　　　　　　　　　　　　　　　（141）
　　8.2　制定成本標準　　　　　　　　　　　　　　　　　　　　　　　（142）
　　8.3　基本的差異分析　　　　　　　　　　　　　　　　　　　　　　（146）
　　8.4　銷售差異　　　　　　　　　　　　　　　　　　　　　　　　　（155）
　　8.5　標準完全成本法下的預算利潤和實際利潤的調節　　　　　　　　（157）
　　8.6　標準變動成本法下的預算利潤（或邊際貢獻）和實際利潤（或邊際貢獻）的
　　　　調節　　　　　　　　　　　　　　　　　　　　　　　　　　　　（157）

第9章　貨幣的時間價值和資本投資評價　　　　　　　　　　　　　　　（163）
　　9.1　資本投資和資本性支出預算　　　　　　　　　　　　　　　　　（163）
　　9.2　貨幣的時間價值　　　　　　　　　　　　　　　　　　　　　　（165）
　　9.3　資本投資評價的非貼現現金流量模型　　　　　　　　　　　　　（174）
　　9.4　資本投資評價的貼現現金流量模型　　　　　　　　　　　　　　（178）
　　9.5　資本投資評估模型的比較　　　　　　　　　　　　　　　　　　（186）

第10章　業績評價　　　　　　　　　　　　　　　　　　　　　　　　　（190）
　　10.1　績效評價概述　　　　　　　　　　　　　　　　　　　　　　　（190）
　　10.2　經濟組織的業績評價　　　　　　　　　　　　　　　　　　　　（192）
　　10.3　責任中心的業績評價　　　　　　　　　　　　　　　　　　　　（196）
　　10.4　經濟增加值　　　　　　　　　　　　　　　　　　　　　　　　（198）
　　10.5　平衡計分卡　　　　　　　　　　　　　　　　　　　　　　　　（201）

附錄：現值系數和終值系數表　　　　　　　　　　　　　　　　　　　　（208）

Chapter 1　Management Accounting Overview

Learning Objectives

After the study of this chapter, you should be able to:
a) Describe the purpose and role of cost and management accounting within an organisation.
b) Compare and contrast financial accounting with cost and management accounting. Explain and illustrate production and non-production costs.
c) Describe the different elements of production cost-materials, labour and overheads.
d) Describe and illustrate, graphically, different types of cost behaviour.
e) Explain and illustrate the concept of cost objects, cost units and cost centre.
f) Distinguish between cost, profit, investment and revenue centre.

1.1　Management Accounting, Financial Accounting and Cost Accounting

1.1.1　Financial Accounting

　　財務會計是一種通過記錄、匯總和報告經濟業務,為企業外部用戶(股東、債權人等企業外部的人)提供財務訊息而進行的經濟管理活動。

Financial accounting is the process of recording, summarizing and reporting the myriad of transactions from a business, so as to provide information about financial position for external users (stockholders, creditors, and others who are outside an organization). Courses in financial accounting cover the generally accepted accounting principles which must be followed when reporting the results of a corporation's past transactions on its balance sheet, income statement, statement of cash flows, and statement of changes in stockholders' equity.

1.1.2　Management Accounting

　　管理會計重在向企業內部提供訊息,使管理者可以更有效地經營公司。管理會計通過計量、分析、報告財務以及非財務訊息,幫助管理者做出決策,以實現組織的目標。

Managerial accounting has its focus on providing information within the company so that its management can operate the company more effectively. Management accounting measures, analyzes, and reports financial and non-financial information that helps managers make decisions to fulfill the goals of an organization. Management accounting information and reports do not have to follow set principles or rules.

1.1.3 Cost Accounting

成本會計為管理會計和財務會計提供了訊息。成本會計計量、分析、報告企業為了獲取或使用資源而產生的與成本相關的財務和非財務信息。

Cost accounting provides information for management accounting and financial accounting. Cost accounting measures, analyzes, and reports financial and non-financial information relating to the costs of acquiring or using resources in an organization.

Comparisons between Cost Accounting and Management Accounting:

a) The scope of management accounting is broader than that of cost accounting.

b) Cost accounting only provides cost information whereas management accounting provides all types of accounting information.

c) Cost accounting focuses on cost ascertainment and cost control whereas management accounting focuses on decision-making.

d) Cost Accounting is a part of Management Accounting whereas Management Accounting is an extension of managerial aspects of cost accounting with the ultimate intention to protect the interests of the business.

1.1.4 Differences Between Management Accounting and Financial Accounting

Most students study management accounting after taking an initial course in financial accounting. These two subjects differ in many ways, please look at table 1.1 as following:

Table 1.1   Contrasting Financial Accounting and Management Accounting

	Financial Accounting	Management Accounting
Audience	External parties; e.g. shareholders and lenders	Internal users: managers and employees
Objective	To disclose the end results of the business, and the financial condition of the business on a particular date	To aid planning, controlling and decision making
Legal requirement	Legally required to prepare financial accounting reports and prescribed by standards such as GAAP	None
Focuses	Focuses on history	Focuses on the present and forecasts for the future
Format	Reported in a specific format	Informal format
Duration	Quarterly and annually	Whenever needed
Information	Monetary, verifiable information	Monetary and company goal driven information

1.2 Management Information

1.2.1 Data and Information

Data is raw, unorganized facts that need to be processed. Data can be something simple and seemingly random and useless until it is organized. For instance, each student's test score is one piece of data.

Information is processed data, which is meaningful and useful to the users of information. For instance, the average score of a school is information that can be derived from the given data, please look at figure 1.1 as following.

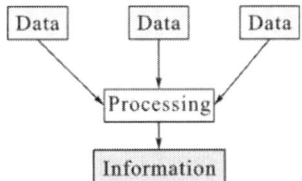

Figure 1.1   Information is Created from Data

1.2.2 The Features of Effective Management Information

Management information is information that is given to the people who is in charge of running an organization.

Information must have certain features and meet certain criteria so that it can be useful to decision maker.

The features of effective management information can be defined by 「ACCURATE」 acronym as follows:

(1) A-Accurate

Accurate information helps businesses make the correct decisions. For example, if a sales department sent wrong sales revenue figures to the finance department, this might result in incorrect tax calculations which would leads to law-breaking event.

(2) C-Complete

It should contain all the facts that are necessary for the decision maker. Nothing important should be left out. For example, managers sometimes only want to know something out-of-ordinary. In this case, a complete report called reporting by exception would be useful.

(3) C-Cost Effective

The information will not be desirable if the cost of obtaining is more than the benefits it expected to provide. The cost of gathering data and processing it into information must be weighed against the benefits derived from using such information.

(4) U-Understandable
Information needs to be understandable by the users. If a non-financial manager needed information about net-income for the last 5 years, financial statement would be a set of complex figures without clear totals in front of him then the non-financial may get confused. So accountants should always be careful about presentation of information to non-financial managers.

(5) R-Relevant
Relevant information is information that is directly related to the business organization's need. It is relatively meaningless to provide a manager with information that he does not know why he has been given.

(6) A-Accessible
Information should be accessible via the right channel and to the right person. Advances in technology have made information more accessible today than ever before. For example, information can be delivered by e-mail, electronic file transfer, and so on.

(7) T-Timely
Information must be delivered at the right time. For example, if a Manager needs to check the sales figures for today, then accountant should provide information from a cash register immediately at any time during the day.

(8) E-Easy
Information should be easy to use.

1.2.3 The Managerial Processes of Planning, Decision-Making and Control

Management Process refers to the activity which involves Planning, Controlling and Decision Making.

Planning involves setting objectives, searching for alternative courses of action and gathering data about alternatives. In other words, planning requires clear objectives and identification of method to achieve those objectives.

There are three levels of planning (also called 「planning horizons」). The planning horizon is the length of time an organization will look into the future.

a) Strategic planning refers to senior managers set long-term goals and plan approximately 3 to 5 years ahead.

b) Tactical planning refers to senior managers set short-term goals and plan approximately 1 year ahead.

c) Operational planning refers to all managers including junior managers are involved in making daily decisions.

1.2.4 Responsibility Centers

Responsibility accounting(責任會計) is a reporting system that compiles revenue, cost, and profit information at the level of those individual managers most directly responsible for them. The intent is to provide this information to those people most able to act upon it, as well as to judge their performance. Responsibility accounting is most commonly used in an organization that distributes responsibility down through the corporate hierarchy.

Responsibility center (責任中心) is a segment of a business where an individual manager is held responsible for the segment's performance.

There are three types of responsibility centers, listed at table 1.2:

Table 1.2　　　　　**Three Types of Responsibility Centers**

1	Cost Centers	Are units in which the function is to incur costs
2	Profit Centers	Have characteristics of cost centers, but also must generate profit
3	Investment Centers	Must concern itself with the overall returns on investments under its purview
4	Revenue Centers	Have authority over sales only and have very little control over costs

This approach allows responsibility to be assigned to the segment managers that have the greatest amount of influence over the key elements to be managed. These elements include revenue for a revenue center (a segment that mainly generates revenue with relatively little costs), costs for a cost center (a segment that generates costs, but no revenue), a measure of profitability for a profit center (a segment that generates both revenue and costs) and return on investment (ROI) for an investment center (a segment such as a division of a company where the manager controls the acquisition and utilization of assets, as well as revenue and costs).

1.3 Cost and Cost Classification

1.3.1 The Types of Cost Classification

One of the purposes of management accounting is to provide managers with information about the costs of products or services. Companies incur different types of costs that can be classified based on certain characteristics.

By behavior. Costs can be classified as fixed costs, variable costs, semi-variable costs and stepped fixed cost.

By element. For example, materials, labor, and expenses (overhead).

By traceability. Direct costs are closely related and traceable to each item produced. Indirect costs are not so easy to relate and trace to each unit of production.

By function. For example, costs related to research development, marketing, training, and manufacturing.

1.3.2 Product Costs and Period Costs

Product costs are costs assigned to the manufacture of products and recognized for financial reporting when sold. Since the matching principle of accounting requires expenses to be matched to the revenue they generate, therefore it is necessary to expense product costs only when the revenue from the sale of products is realized. This is achieved by debiting product costs to the cost of goods manufactured and thus expensed only at the time of sale of such goods.

Examples of product costs are direct materials, direct labor, factory wages, factory depreciation, etc.

Period costs are on the other hand are all costs other than product costs. These costs are not incurred on the manufacturing process and therefore these cannot be assigned to cost of goods manufactured. Period costs are thus expensed in the period in which they are incurred.

Example of period costs are advertising, sales commissions, office supplies, office depreciation, research and development costs.

Period costs can be further classified into selling costs and administrative costs.

1.3.3 Direct and Indirect Costs

(1) Direct Costs

Direct costs(直接成本) of a cost object are related to the particular cost object and can be traced to it in an economically feasible (cost-effective) way. Direct costs include direct materials, direct labor and direct expenses.

Direct materials. For example, steel for making cars.

Direct labour. For example, the wages of the workers on the car production line.

Direct expenses. For example, the cost of maintaining the machine used to produce the cars.

The total of direct costs is also called the prime cost.

(2) Indirect Costs

Indirect costs(間接成本) of a cost object are related to the particular cost object but cannot be traced to it in an economically feasible (cost-effective) way. Indirect costs include indirect materials, indirect labor and indirect expenses.

Indirect materials. For example, lubricating oil.

Indirect labour. For example, the salaries of administrators who oversee production of many different types of cars produced.

Indirect expenses. For example, the rent expenses of car factory.

The total of indirect costs is also called overheads.

1.3.4 Fixed and Variable Costs

成本習性也稱為成本性態,指成本的變動與業務量之間的依存關係。

(1) Cost Behavior

Cost can be classified by behavior. Cost behavior is the way that costs change when there is a change in an organization's level of activity. Activity level refers to the amount of work done or the volume of production.

There are four types of costs that are classified on the basis of behavior.

(2) Variable Costs

變動成本是指其總額隨著業務量成正比例變動的那部分成本。

Variable costs(變動成本) are directly proportional to the level of activity. If the number of units sold increases by 10% then variable selling and distribution costs would increase by 10% also. Variable cost per unit keeps constant when activity level changes.

Total VC — Increase when activity level increases
— Decrease when activity level decreases
VC/u — Constant when activity level changes

But the variable cost per unit can differ in various relevant ranges of activity levels.

Examples of variable costs include:

a) The cost of raw materials, Where there is no discount for bulk purchasing since bulk purchase discounts reduce the cost of purchases.

b) Direct labor costs are, for very important reasons, classed as a variable cost even though basic wages are usually fixed.

c) Sales commission is variable in relation to the volume or value of sales.

d) Bonus payment for productivity to employees might be variable once a certain level of output is achieved.

On a graph, variable costs are shown in figure 1.2:

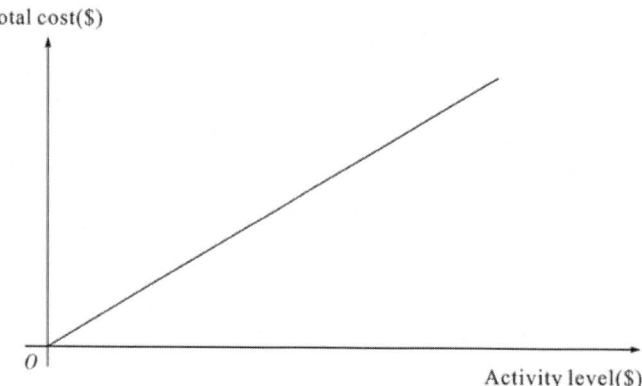

Figure 1.2　Variable Costs

(3) Fixed Costs

固定成本是指其總額在一定時期和一定業務量範圍內不隨業務量發生任何變動的那部分成本。每單位的固定成本隨著業務量的增加而減少。

Fixed costs (固定成本) are not affected in total by the level of activity, while fixed cost per unit decreases when activity level increases.

  Total FC — Constant when activity level changes
    FC/u — Increase when activity level decreases
      — Decrease when activity level increases

An example would be the rent of a single factory building. No matter how many units are made, the rent is fixed.

Other examples of fixed costs are as follows:

a) The salary of the managing director (per month or per annum).

b) Straight line depreciation of a single machine (per month or per annum).

On a graph, fixed costs are shown in figure 1.3:

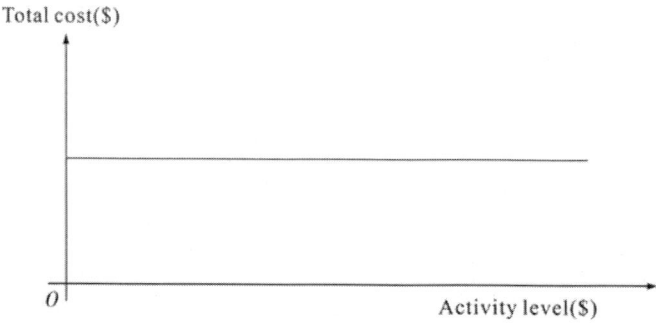

Figure 1.3　Fixed Costs

階梯成本隨產量的變化而呈階梯型增長,產量在一定限度內,這種成本不變,而當產量增長到一定限度後,這種成本就跳躍到一個新水平。

(4) Stepped Fixed Costs

Stepped fixed costs(階梯成本) keep constant for a range of activity level, and then change, and keep constant again for another range.

An example would be the salary of supervisors, one supervisor for up to ten workers, two for up to twenty workers, etc. Other examples of these costs include the following:

a) Depreciation.
b) Rent.
c) Basic pay of employee.
d) Royalties.

On a graph, stepped fixed costs are shown in figure 1.4:

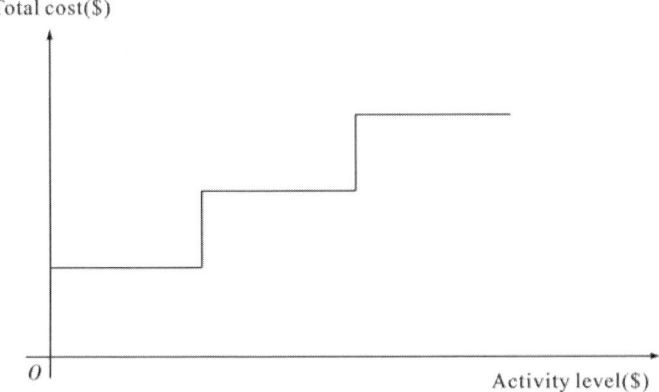

Figure 1.4 Stepped Fixed Costs

半變動成本當中既包含固定成本的部分又包含變動成本的部分。

(5) Semi-variable Costs

Semi-variable costs(半變動成本) are cost which contain both fixed and variable elements and so are partly affected by changes in the level of activity.

  Total VC — Increase when activity level increases
  Total FC — Constant when activity level changes
  Total Semi-VC — Increase when activity level increases
  VC/u — Constant when activity level changes
  FC/u — Decrease when activity level increases
  Semi-VC/u — Decrease when activity level increases

Examples of these costs include the followings:

a) Electricity and gas bills.
 (i) Fixed cost = standing charge
 (ii) Variable cost = charge per unit of electricity used
b) Salesman's salary.
 (i) Fixed cost = basic salary

(ii) Variable cost = commission on sales made
c) Costs of running a car.
(i) Fixed cost = road tax, insurance
(ii) Variable cost = petrol, oil, repairs (which vary with mile traveled)

On a graph, Semi-variable costs are shown in figure 1.5:

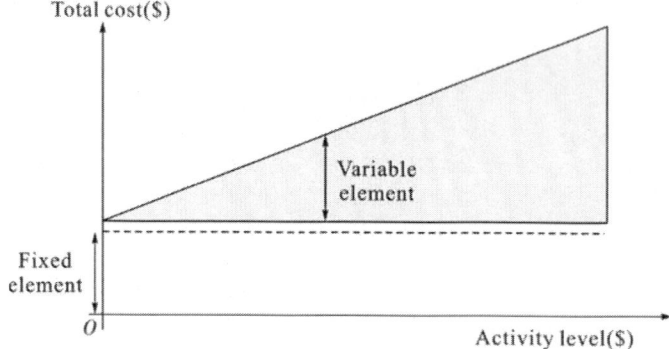

Figure 1.5 Semi-variable Costs

1.3.5 Cost Objects, Cost Units and Cost Centres

(1) Cost Objects

A cost object is a term used to describe something to which costs are assigned. For example, we might want to know the cost of making one unit of product. In this case the cost object is one unit of product. If we want to know the cost of operating a department, then the cost object is the department or factory.

(2) Cost Units

Cost unit is a quantity or unit of a product or service whose cost is computed, used as a standard for comparison with other costs.

There are some types of cost objects:
Output. The most common cost objects are a company's products and services, since it wants to know the cost of its output for profitability analysis and price setting.

Operational. A cost object can be within a company, such as a department, machining operation, production line, or process. For example, you could track the cost of designing a new product, or a customer service call, or of reworking a returned product.

Business relationship. A cost object can be outside of a company – there may be a need to accumulate costs for a supplier or a customer, to determine the cost of dealing with that entity. Another variation on the concept is the cost of renewing a license with a government agency.

It may be necessary to have a cost object in order to derive pricing from a baseline cost, or to see if costs are reasonable, or to derive the full cost of a relationship with another entity.

(3) Cost Centers

A cost center is a business unit that is only responsible for the costs that it incurs. The manager of a cost center is not responsible for revenue generation or asset usage. The performance of a cost center is usually evaluated through the comparison of budgeted to actual costs. The costs incurred by a cost center may be aggregated into a cost pool and allocated to other business units, if the cost center performs services for the other business units.

A cost centre could be:
Accounting department
Human resources department
IT department
Maintenance department
Research & development

The main function of a cost center is to track expenses. The staff of a cost center is only responsible for the costs and does not bear any responsibility regarding revenue or investment decisions. Expense segmentation into cost centers allows for greater control of total costs. Accounting for resources at a finer level such as a cost center allows for more accurate forecasts and calculations based on future changes.

QUESTIONS:

1. The following statements relate to financial accounting or to cost and management accounting:
(i) Financial accounts are historical records.
(ii) Cost accounting is part of financial accounting and establishes costs incurred by an organisation.
(iii) Management accounting is used to aid planning, control and decision making.
Which of the statements are correct?
A. (i) and (ii) only
B. (i) and (iii) only
C. (ii) and (iii) only
D. (i), (ii) and (iii)

2. Which of the following is the manager of a profit centre responsible for?
(i) Revenues of the centre.
(ii) Costs of the centre.
(iii) Assets employed in the centre.
A. (i) only.
B. (ii) only.
C. (i) and (ii) only.
D. (i), (ii) and (iii).

3. Which of the following is usually classed as a step cost?
A. Supervisor's wages
B. Raw materials
C. Rates
D. Telephone

4. All of the following may be cost objects except.
A. A cost centre
B. A customer
C. A manager
D. A product

5. The diagram represents the behaviour of a cost item as the level of output changes:

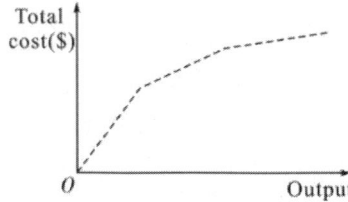

Figure 1.6　Outup Changes

Which one of the following situations is described by the figure 1.6?

A. Discounts are received on additional purchases of material when certain quantities are purchased

B. Employees are paid a guaranteed weekly wage, together with bonuses for higher levels of production

C. A licence is purchased from the government that allows unlimited production

D. Additional space is rented to cope with the need to increase production

6. All of the followings are classified as product costs except:

A. Factory depreciation

B. Machine insurance

C. Factory wages

D. Research and development costs

7. The prime cost of a product is the sum of the labour and materials costs that are identifiable to individual units of the product.

This statement is:

A. True

B. False

8. A manufacturing firm is very busy and overtime is being worked.

The amount of overtime premium contained in direct wages would normally be classed as:

A. Part of prime cost

B. Factory overheads

C. Direct labour costs

D. Administrative overheads

9. The total materials cost of a company is such that when total purchases exceed 15,000 units in any period, then all units purchased, including the first 15,000, are invoiced at a lower cost per unit.

Determine which of the following graphical representations is consistent with the behaviour of the total materials cost in a period.

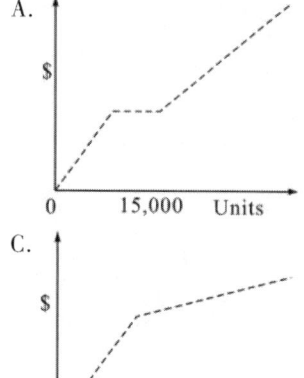
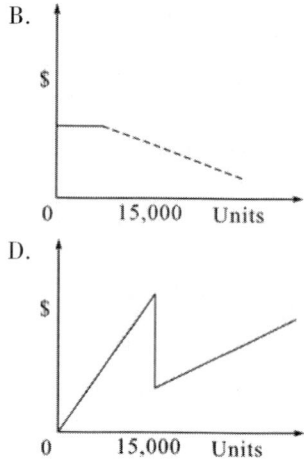

10. Which of the following are true of marginal costing?

(i) The marginal cost of a product includes an allowance for fixed production costs.

(ii) The marginal cost of a product represents the additional cost of producing an extra unit.

(iii) If the inventory increases over a year, the profits under absorption costing will be lower than with marginal costing.

A. (i) only.

B. (ii) only.

C. (ii) and (iii) only.

D. (i), (ii) and (iii).

Chapter 2　Cost Accounting Principles

Learning Objectives

After the study of this chapter, you should be able to:
a) Describe the different procedures and documents necessary for the ordering, receiving and issuing of materials from inventory.
b) Identify, explain and calculate the cost of the ordering and holding inventory (including buffer inventory).
c) Calculate and interpret optimal reorder quantities. Calculate and interpret optimal reorder quantities when discounts apply.
d) Calculate the value of closing inventory and material issues using LIFO, FIFO and average methods.
e) Calculate direct and indirect costs of labour. Describe different remuneration methods: time–based systems, piecework systems and individual and group incentive schemes.
f) Calculate and interpret labour efficiency, capacity and production volume ratios.
g) Allocate and apportion production overheads to cost centres using an appropriate basis. Re-apportion service cost centre costs to production cost centres (using the reciprocal method where service cost centres work for each other). Select, apply and discuss appropriate bases for absorption rates. Prepare journal and ledger entries for manufacturing overheads incurred and absorbed. Calculate and explain the under and over absorption of overheads.

2.1　Accounting for Material

2.1.1　Direct Material and Indirect Material

Direct materials are any raw materials, parts or sub-part required for the completion of a product. The cost can be directly traced and attributable to a specific product or job.

Examples of indirect materials are: steel used in production of car, wood used in production of furniture, flour used for producing bread, etc.

Indirect materials are materials used in the production process, but which cannot be associated to a specific product or job.

Examples of indirect materials are: oil, glue, tape, etc.

2.1.2 Issuing Inventory

The accounting procedures for issuing inventory are as follows:

(1) Materials Requisition Notes

In a paper-based system, the materials requisition is prepared by production manager or department supervisor. When the storeroom keeper receives a properly signed requisition, the requisitioned materials are released, please look at figure 2.1 as following:

```
┌─────────────────────────────────────────────────────────┐
│  SEMINOLE        MATERIALS REQUISITION                  │
│  MFG                                                    │
│                                                         │
│  Date  January 19, 2002              No.  632           │
│                                                         │
│  To:   D. Graham                                        │
│                                                         │
│  ┌──────────┬──────────────────┬────────────┬─────────┐ │
│  │ QUANTITY │   DESCRIPTION    │ UNIT PRICE │ AMOUNT  │ │
│  ├──────────┼──────────────────┼────────────┼─────────┤ │
│  │ 100 Gal. │ Adhesive Compound│  $1.565    │ $156 50 │ │
│  │          │ Grade A102       │            │         │ │
│  └──────────┴──────────────────┴────────────┴─────────┘ │
│                                                         │
│  Approved by  E.B.         Issued by  B.W.              │
│  Received by  L.M.                                      │
│                                                         │
│  Charged to Job/Dept. 300  Factory Overhead Expense Account ___ │
└─────────────────────────────────────────────────────────┘
```

Figure 2.1 Material Requistion

(2) Materials Returned Notes

After materials are requisitioned, occasionally some unused materials must be returned to the storeroom. Materials returned notes are used to record the reason for return, accompanying materials to storeroom.

Chapter 2   Cost Accounting Principles   17

(3) **Materials Transfer Notes**
These documents are used to record the transfer of materials from one production department to another.

2.1.3 Accounting for Inventory

(1) **Material Inventory Account**
Accounting transaction relating to materials is recorded in the material inventory account.
Debit entries reflect an increase in inventory. Credit entries reflect a decrease in inventory.

(2) **Inventory Valuation**
The accounting method that a company decides to use to determine its inventory costs can directly impact the closing inventory value and values of issues from stores. Three inventory-costing methods are widely used by both public and private companies:
a) First-In-First-Out Method (FIFO). According to FIFO, it is assumed that materials are issued in the order in which they were delivered into inventory. Closing stock will be valued at the price of the latest goods remaining in stock. When material prices increased, the cost of production would be lower than costs under other methods and closing inventory values would be higher than values under other methods. The reverse would be the case when prices were falling.

先進先出法是指根據先入庫先發出的原則,對發出的存貨以先入庫存貨的單價計算發出存貨成本的方法。

Objectives and Advantages of FIFO Method:
One objective of FIFO is to approximate the physical flow of goods. When the physical flow of goods is actually first-in-first-out, the FIFO method closely approximates specific identification. At the same time, it does not permit manipulation of income because the enterprise is not free to pick a certain cost item to be charged to expense.
Another advantage of the FIFO method is that the ending inventory is close to current cost. Because the fist goods in are the first goods out, the ending inventory amount will be composed of the most recent purchases. This is particularly true where the inventory turnover is rapid. This approach generally provides a reasonable approximation of replacement cost on the balance sheet when price changes have not occurred since the most recent purchases.

Disadvantages of FIFO Method:

The basic disadvantages of first in first out method (FIFO Method) are that costs are not matched against current revenues on the income statement. The oldest costs are charged against the more revenue, which can lead to distortion in gross profit and net income.

後進先出法下,期末存貨按最早發生的成本計價,銷貨成本按最近發生的成本計價。

b) Last-In-First-Out Method (LIFO). Materials are issued at the price of the most recently received goods. Closing inventory will be valued at the price of the earliest goods remaining in stock.

Advantages of Last-In-First-Out (LIFO) Method:

Matching: In LIFO, the more recent costs are matched against current revenues to provide a better measure of current revenues.

Tax Benefits/Improved Cash Flow: Tax benefits are the major reason why LIFO has become popular. As long as the price level increases and inventory quantities do not decrease, a deferral of income tax occurs, because the items most recently purchased at the higher price level are matched against revenues.

Future Earnings Hedge: With LIFO, a company's future reported earnings will not be affected substantially by future price declines. LIFO eliminates or substantially minimizes write-downs to market as a result of price decreases. Since the most recent inventory is sold first, there is not much ending inventory sitting around at high prices vulnerable to a price decline. In contrast, inventory costed under FIFO is more vulnerable to price decline, which can reduce net income substantially.

Despite its advantages, LIFO has the following drawbacks:

Inventory Understated. LIFO may have a distorting effect on the company's balance sheet. The inventory valuation is normally outdated because the oldest costs remain in the inventory. This understatement makes the working capital position of the company appear worse than it really is.

Physical Flow. LIFO does not approximate the physical flow of the items.

移動加權平均法是指以每次進貨的成本加上原有庫存存貨的成本,除以每次進

c) Moving-Average (Unit) Cost. Moving-Average (Unit) Cost is a method of calculating ending Inventory cost. Assume that both beginning inventory and beginning inventory cost are known. From them the cost per unit of beginning

貨數量與原有庫存存貨的數量之和，據以計算加權平均單位成本，以此為基礎計算當月發出存貨的成本和期末存貨的成本的一種方法。

Inventory can be calculated. During the year, multiple purchases are made. Each time, purchase costs are added to beginning inventory cost to get cost of current Inventory. Similarly, the number of units bought is added to beginning inventory to get current goods available for sale. After each purchase, cost of current inventory is divided by current goods available for sale to get current cost per unit on goods. Also during the year, multiple sales happen. The current goods available for sale is deducted by the amount of goods sold, and the cost of current inventory is deducted by the amount of goods sold times the latest (before this sale) current cost per unit on goods. This deducted amount is added to cost of goods sold. At the end of the year, the last cost per unit on goods, along with a physical count, is used to determine ending inventory cost.

2.1.4 Stock Movement and Stocktaking

(1) Perpetual Inventory System

Under the perpetual inventory system, an entity continually updates its inventory records to account for additions and subtractions in inventory for activities such as: goods sold from stock, received inventory items, and so on.

Inventory records are updated usually using bin cards and stores ledger cards.

The store-keeper uses bin cards to record daily receipt, issues and balance of each item of stores.

A typical bin card is shown in table 2.1:

Table 2.1   **Bin card**

Name of Material .... Size and .... Accounting Unit ....
Code No .... Specification .... Location ....

Date	Particulars Reference to Goods Received Note or Material Requisition	Receipt	Issue	Balance	Remarks

The quantity information on the bin cards is also entered on a document called stores ledger cards. Please look at figure 2.2 as following:

Table 2.2   **STOCK LEDGER CARD**

Date	Details	Purchases			Cost of Goods Sold			Balance		
		Qty	Unit cost $	Total cost $	Qty	Unit cost $	Total cost $	Qty	Unit cost $	Total cost $

(2) Stocktaking

Stocktaking is simply physical checking or counting of inventory held by the entity. The main objective of stocktaking is to identify stock discrepancies and relevant correction adjustments.

Under periodic stocktaking, stock is counted once at the end of period usually a year. During the stocktaking, all other operations had to be put on hold to avoid disruption to the stocktaking.

Under continuous stocktaking, stocks are counted on an 「on-going basis」. Each item of stock is checked at least once during the year, but many items, particularly expensive items or rapidly moving items will be checked more frequently. This method tends to avoid the disruption of the periodic method.

2.1.5 The Cost of Holding Inventory

(1) Reasons for Holding Inventory

The reasons that an entity holds inventory include: taking advantage of bulk purchasing discounts, meeting expected demand, etc.

(2) Stock Costs

a) Ordering Costs. Ordering costs are costs that are incurred on obtaining additional inventories. They include costs incurred on communicating the order.

For example, administrative costs, transport costs, and production run costs.

b) Holding Costs. Holding costs represent the costs incurred on holding inventory in hand.

For example, costs of storage, insurance costs, risk of obsolescence.

c) Stock-Out-Costs. Stock-Out-Costs are costs associated with running out of inventory.

For example, losses, compensation to customers, idle time costs.

(3) Inventory Control Level

a) Free Stock. Free stock is the quantity of stock available to the company.

Free stock = (P+Q) - U

P = Physical stock (stock in hand)

Q = Outstanding orders with suppliers (stock on order)

U = Unfulfilled requirements (stock allocated)

Example 2.1

There are 350 units of material A in stock. An order for 680 is expected to be received and a material requisition for 400 has not been issued to the production cost centre. Required: What is free stock?
Solution:
Free stock = (P+Q)−U = (350+680)−400 = 630 units

b) Reorder Level. Reorder level refers to the level of free stock at which an order should be placed for replacement inventory.
Reorder level = Maximum usage × Maximum lead time
Lead time is the time that elapses between placing and receiving an order.

Example 2.2

Maximum usage = 700kg/day
Lead time = 4~6days
Required:
Calculate the reorder level.
Solution:
Reorder level = Maximum usage × Maximum lead time
   = 700kg/day×6days
   = 4,200kg

c) Minimum Level. Minimum level of inventory is a warning that inventory levels are dangerously low. If inventory falls below this level, emergency action to replenish is required to prevent stock-out.
Minimum level
   = Reorder level − (Average usage × Average lead time)

d) Maximum Level. Maximum level of inventory is a warning that inventory levels are dangerously high. Maximum level is set for control purposes, which actual stock-holding should not exceed.
Maximum level
   = Reorder level + Reorder quantity − (Minimum usage × Minimum lead time)

e) Economic Order Quantity(經濟訂貨批量). The Economic Order Quantity (EOQ) is the number of units that a company should add to inventory with each order to minimize the total costs of inventory—such as ordering costs and

當企業按照經濟訂貨批量來訂貨時，可實現訂貨成本和儲存成本之和最小化。

holding costs. Ordering costs and holding costs are quite opposite to each other. If we need to minimize holding costs we have to place small order which increases the ordering costs. If we want minimize our ordering costs we have to place few orders in a year and this requires placing large orders which in turn increases the total holding costs for the period.

$$EOQ = \sqrt{\frac{2 \times Demand \times CostPerOrder}{HoldingCostPerUnit}}$$

EOQ assumptions:
- no lead time
- no discount
- Even demand throughout the year

Example 2.3

ABC Ltd. is engaged in sale of footballs. Its cost per order is $400 and its carrying cost unit is $10 per unit per annum. The company has a demand for 20,000 units per year. Calculate the order size, total orders required during a year, total carrying cost and total ordering cost for the year.

Solution:

$$EOQ = \sqrt{\frac{2 \times 20,000 \times 400}{10}} \approx 1,265 \text{units}$$

Annual demand is 20,000 units so the company will have to place 16 orders (= annual demand of 20,000 divided by order size of 1,265). Total ordering cost is hence $64,000 ($400 multiplied by 16).

Average inventory held is 632.5 i.e. [(0 + 1,265)/2] which means total carrying costs of $6,325 (i.e. 632.5 × $10).

2.2　Accounting for Labor

2.2.1　Direct and Indirect Labor

For costing purposes, labor can be classified into two broad categories: direct labor and indirect labor. The distinction between direct and indirect labor is important because it helps:

a) To determine accurate product cost.
b) To measure efficiency of performance.
c) To minimize error in overhead allocation.
d) To ensure better cost analysis for decision-making and control.

Direct labor cost is a part of payroll that can be specifically assigned to the manufacture of a product or provision of a service. When a business manufactures products, direct labor is considered to be the labor of the production crew that produces goods, such as machine operators, software programmers, assembly line operators, and so forth. When a business provides services, direct labor is considered to be the labor of those people who provide services directly to customers, such as nurses and lawyers.

Indirect labor is the cost of any labor that do not directly produce goods or services, but who make their production possible or more efficient. Indirect labor costs are not readily identifiable with a specific task or work order.

Examples of indirect labor include: accountants, quality control staff, production supervisors, workshop cleaner, security guards, and so forth. Wages and salaries paid to such staff are treated as indirect labour cost which is included in overheads.

Indirect labor costs also include the following:
- Overtime premium (not required by customer)

Overtime premium is generally treated as an indirect cost. Exception: If customer requests that overtime is worked in order to complete a job as soon as possible, then overtime premium will be treated as direct cost.

-Idle Time

Idle time is an indirect labor cost. This is the non-productive time (during which an employee is still paid) of employees or machines, or both, due to machine breakdowns, production scheduling Problems, material shortages. It is also called waiting time, allowed time, or downtime.

2.2.2 Calculating Labor in Productions and Services

Method of Recording Time spent doing jobs.
Methods can include:
(1) **Clock Cards**
A clock card is a document which records the starting and finishing time for an employee.
(2) **Job Sheets**
A job sheet records the number of each type of product that an employee has produced during a period.
(3) **Time Sheets**
A time sheet is a method for recording the amount of a

worker's time spent on each job. Traditionally a sheet of paper with the data arranged in tabular format, a time sheet is now often a digital document or spreadsheet.

An example of weekly Time-sheet is illustrated on table 2.3:

Table 2.3　　　　　　　　　Weekly Employee Timesheet

Weekly Employee Timesheet

Timesheets by Vertex42.com　　　　　　　　　　　　　　　　© 2008 Vertex42 LLC

[Company Name]

[Address 1]
[Address 2]
[City, State ZIP]
[Phone]

Employee Name: _____

Supervisor Name: _____

Week of: 10/6/2008

Day of Week	Regular Hrs	Overtime Hrs	Sick	Vacation	Holiday	Unpaid Leave	Other	TOTAL Hrs
Mon 10/6								0.00
Tue 10/7	8.00	0.43						8.43
Wed 10/8								0.00
Thu 10/9								0.00
Fri 10/10								0.00
Sat 10/11								0.00
Sun 10/12								0.00
Total Hrs:	8.00	0.43	0.00	0.00	0.00	0.00	0.00	8.43
Rate/Hour:	15.00	23.00	15.00	15.00	15.00	0.00	0.00	
Total Pay:	120.00	9.89	0.00	0.00	0.00	0.00	0.00	$ 129.89

Total Hours Reported:　8.43
Total Pay:　123.83

Employee Signature　　　　　Date

Supervisor Signature　　　　Date

2.2.3 Accounting for Labor Cost

Labor costs are expenses and recorded in an organization's income statement. Accounting transactions relating to labor are recorded in the labor account.

The gross pay is debited to the labor account and the direct cost element is then transferred to the work-in-progress account whilst the indirect cost element is transferred to the production overhead account.

2.2.4 Labor Remuneration Method

Labor remuneration methods refer to the ways how employees are being paid based on their working performance. There are two ways to calculate remuneration: time-based and output-based.

(1) Time-based Remuneration (計時工資)

According to time-based remuneration system, employees will be paid on a period of time worked. It can be hourly, daily, weekly, monthly or others.

Labor pay is divided into normal time pay and overtime pay. Normal time pay is the pay which received from attending to normal working hours. While overtime pay is the pay which received for overtime hours worked.

The formula for Time-based remuneration is as follows.
Labor remuneration = (Hours worked × Basic rate of pay per hour) + (Overtime hours worked × Overtime premium per hour)

Example 2.4

Mike works as a software programmer in a company. Normal working time is 8-hour day for 5 days week. Mike worked 50 hours during a particular week. Basic rate is $10 per hour. Overtime pay is at 1.75 of basic.
Calculate the total labor pay to Mike.

Solution:
Normal time pay (8hrs × 5days × $10/hr) = $400
Overtime pay (10hrs × $10/hr × 1.75) = $175
Total labor pay = $575

(2) Output Remuneration (計件工資)

Instead of calculating labor remuneration based on time, another approach is based on output. The formula for Output-Based remuneration is as follows:
Labor remuneration = (No. of units × Rate per unit)
Output based remuneration sometimes is called piecework rates. There are two main categories of rates:
Straight piecework system: The wages of the worker depend upon his output and rate of each unit of output; it is in fact independent of the time taken by him.
Differential piece work system: This system provides for higher rewards to more efficient workers. For different levels

of output below and above the standard, different piece rates are applicable.

Example 2.5

A company operates a piecework system with a guaranteed earning as follows:

Up to 50 units　　　　$0.6 per unit
51~70 units　　　　　$0.7 per unit
71 and above　　　　 $0.8 per unit

All workers are guaranteed a pay of $40 per day regardless of output achieved.

	Jacky	Tommy	Judy
Production (units)	55	65	75

Calculate the gross wages for each employee.

Solution:
The gross wages for Jacky = 50×0.6+5×0.7 = $33.5
This is less than the guaranteed minimum therefore Jacky would be paid $40.
the gross wages for Tommy = 50×0.6+15×0.7 = $40.5
the gross wages for Judy = 50×0.6+20×0.7+5×0.8 = $48

(3) Bonus or Incentive Schemes(獎金及獎勵機制)

Incentive schemes are introduced in order to increase production. Wage incentive plans may be of two categories: Individual Incentive Plans and Group Incentive Plans.

Under individual incentive plans, remuneration can be measured by the performance of the individual worker. In the case of the group incentive scheme earnings can be measured on the basis of the productivity of the group of workers or entire work force of the organization. The followings are some important individual incentive plans:

a) Halsey Premium Plan. This Plan was developed by F. A. Halsey. This plan assumes that standard time is fixed for each job or operation on the basis of past performance. If a worker completes his job within or more than the standard time, then he will be paid a guaranteed time wage. If a worker finishes his job within or less than the standard time, then he will receive 50% of the time saved plus normal earnings.

The calculation of total earnings under Halsey Premium Plan is:

Total Earning = T×R + 50% (S−T)×R

Where
T – Time Taken
R – Hourly Rate
S – Standard Time

Example 2.6

Calculate the total earnings of the worker using Halsey Premium Plans:
Standard Time: 12 hours
Hourly Rate : $4 per hour
Time Taken 10 hours

Solution:
Earnings using Halsey Premium Plan
Total Earning = T×R+ 50% (S−T) ×R
            = 10×4+ 50% (12−10) ×4 = $44

b) Rowan Plan: This plan was introduced by James Rowan. Under this system, bonus is determined as the proportion of the time taken which the time saved bears to the standard time allowed. Under this system
The calculation of total earnings under Rowan Plan is:
Total Earnings = T×R+ (time saved/S) ×T×R
Where
T – Time Taken
R – Hourly Rate
S – Standard Time
Time saved− Standard Time− Time Taken

Example 2.7

Calculate the total earnings of the worker using Rowan Plan:
Standard Time = 10 hours
Time Taken = 8 hours
Rate per hour = $3 per hour

Solution:
Earnings using Halsey Premium Plan
Total Earnings = T×R+ (time saved/S) ×T×R
             = 8×3+ (2/10) ×8×3
             = $28.8

2.2.5 Labor Turnover

Employee retention is the ability of a firm to convince its employees to remain within the business. It is often measured by the labor turnover of a business.

(1) Types of Turnovers

There are four types of turnovers:

Voluntary is the first type of turnover, which occurs when an employee self-willingly makes the decision to leave the organization.

The second type of turnover is Involuntary, this occurs when the employer makes the decision to discharge an employee and the employee unwillingly leaves his or her position.

The third type of turnover is Functional, which occurs when a low performing employee leaves the organization.

The fourth type of turnover is called Dysfunctional, it ouurs when a high performing employee leaves the organization.

Efficient managers will investigate high levels of labour turnover and aim to keep that turnover rate at a minimum.

(2) Calculation of Labor Turnover

Labor turnover is defined as the proportion of a firm's workforce that leaves during the course of a year.

The formula for calculating labor turnover is shown below:
$$[NELDY/(NEBY+NEEY)/2] \times 100\%$$

Where:
NELDY
= Number of Employees who Left During the Year
NEBY
= Number of Employees at the Beginning of the Year
NEEY
= Number of Employees at the End of the Year

Example 2.8

At the start of the year a business had 40 employees, but during the year 6 staff resigned with 2 new hires, thus leaving 32 staff members at the end of the year. What was the labor turnover rate for the year?

Solution:
$\{6 \div [(40+32) \div 2]\} \times 100\% = 16.7\%$.

2.2.6 Labor Efficiency, Capacity and Production Volume Ratios

(1) Labor Efficiency Ratio

Labor is an important cost in business. Therefore, it is necessary for organizations to continually measure the efficiency of labor against preset targets.

The labor efficiency ratio measures the performance of the workforce by comparing the actual time taken to do a job with the expected time.

The formula for calculating the labor efficiency ratio is as follows:

Labor efficiency ratio
= Standard hours for actual production/Actual hours worked expressed as a percentage

(2) Labor Capacity Ratio

The labor capacity ratio measures the number of hours spent actively working as a percentage of the total hours available for work (full capacity or budgeted hours).

The formula for calculating the labor capacity ratio is as follows:

Capacity ratio
= Actual hours worked/Budgeted hours expressed as a percentage

(3) Labor Production Volume Ratio (「Activity」Ratio)

The labor production volume ratio compares the number of hours expected to be worked to produce actual output with the total hours available for work (full capacity or budgeted hours).

The formula for calculating the labor production volume ratio is as follows:

Labor activity ratio
= Standard hours for actual production/Budgeted hours, or Actual output/Budgeted output expressed as a percentage

Example 2.9

ABC Company budgets to make 50,000 units of toy cars in 5,000 hours (each unit is budgeted to take 0.1 hours) in a year. Actual output during the year was 39,000 units which took 4,200 hours to make.

Required:
Calculate the labor efficiency, capacity and production volume ratio.

Solution:
Standard hours for actual production = 3,900
Actual hours worked = 4,200

Budgeted hours = 5,000
Labor efficiency ratio = (3,900÷4,200) × 100% = 92.9%
Capacity ratio = (4,200÷5,000) × 100% = 84%
Production volume ratio = (3,900÷5,000) × 100% = 78%

2.3　Accounting for Overheads

2.3.1　Recap of Direct and Indirect Expenses

Direct expenses are expenses that can be directly identified with a specific cost centre or cost unit. For instance, the cost of the freight needed to transport goods to and from a manufacturing facility can be directly related to the cost of production.

There are many more types of expenses that are not direct expenses. They are called indirect expenses, because those expenses cannot be directly identified with a specific cost centre or cost unit.

Examples of indirect expenses are:
Facility rent
Facility insurance
Depreciation and amortization
Heat and light

Production overheads are shared out between units of production under「absorption costing」.

2.3.2　Production Overhead Absorption

(1) **Fixed Production Overheads**

Production overheads are the sum of indirect production costs.

Fixed production overheads = Indirect materials + Indirect labor + Indirect expenses

Examples of fixed production overheads include rent, heating and lighting.

One way of recovering fixed production overheads is to calculate the overhead cost allocated to each unit of product.

Example 2.10

Smile Ltd is a manufacturing company producing product S. The Prime cost of product S is $15. Smile Ltd produces and sells 1,100 units in a month.

Smile Ltd estimates its monthly overheads will be:
Heating $3,500
lighting $2,500
Rent $600

Required:
Calculate the overhead cost allocated to each product S and the cost per unit of product S.

Solution:
The overhead cost allocated to each product
= (3,500+2,500+600) ÷ 1,100 = $6
The cost per unit = 6+15 = $21

(2) Absorption Costing

Absorption costing is defined as a method for accumulating the costs associated with a production process and apportioning them to individual products. GAAP (Generally Accepted Accounting Principles) require absorption costing for external reporting.

Absorption costing involves the following stages:
a) Allocation and apportionment.
b) Reapportionment.
c) Absorption.

2.3.3 Allocation and Apportionment

The assignment of cost figures to specific cost objects is a central task in budgeting, planning, and financial reporting. Cost allocation and cost apportionment are methods for attributing cost to particular cost objects.

(1) Allocation

Allocation means the allotment of all items of cost to one particular cost centre. Cost is allocated when the cost centre uses all of the benefits of the expenses and no sharing of the cost is required.

Illustration:
A manufacturing company has two production cost centers, manufacturing and finishing. The company also has two service cost centre, stores and maintenance. The following expenses are expected to be incurred for the next financial year:

Indirect materials—Manufacturing $5,500
                   Finishing $6,500
                   Maintenance $3,000

Indirect labor—　Manufacturing　$10,000
　　　　　　　　Finishing　　　$6,000
　　　　　　　　Stores　　　　 $15,000
　　　　　　　　Maintenance　　$15,500

Required:
Allocate the overheads to the cost centers.

Solution:
(Allocation of overheads is when the entire overhead cost can be traced to one department)
The expenses of production centers and sevice cost centers are listed in table 2.4:

Table 2.4　The Expenses of Production Centers and Sevice Centers

| | Production Cost Centers || Service Cost Centers ||
	Manufacturing	Finishing	Stores	Maintenance
Indirect materials	$5,500	$6,500		$3,000
Indirect labour	$10,000	$6,000	$15,000	$15,500
Total allocated	$15,500	$12,500	$15,000	$18,500

(2) **Apportionment**

Apportionment refers to charge the total value of overhead item between two or more cost centers. It needs a suitable basis for subdivision of cost by cost centers or cost units. Possible bases include the follows:

Rent and rates — area
Heat and light — area
Buildings Insurance — area
Depreciation of machinery — value of machinery

Example 2.11

A manufacturing company has two production departments: Assembly and Finishing. One electricity meter records the usage of both departments and one supervisor is employed to oversee the work of staff in both departments.

The figures budgeted for electricity and supervision expected to be incurred in the forthcoming year, are as follows:

Overhead	Cost
Electricity	$25,000
Supervisor's wages	$60,000
Total overhead charge	$85,000

Some additional information is also available:

	Assembly	Finishing	Total
Number of employees	4	6	10
Floor area (sq m)	1,200	800	2,000

Required:
How should the overhead costs be apportioned?

Solution:
Overhead analysis sheet as showed in table 2.5:

Table 2.5   Overhead Analysis Sheet

Overhead	Basis of apportionment	Assembly	Finishing	Total
Electricity	Floor area (w1)	$15,000	$10,000	$25,000
Supervision	Employees (w2)	$24,000	$36,000	$60,000
	Total	$39,000	$46,000	$85,000

This is called an overhead analysis statement.

Workings:
a) Electricity is apportioned to both departments based on floor area occupied.
Total electricity = $25,000
Total floor area occupied = 2,000 sq m
Apportioned to Assembly department
= 1,200 ÷ 2,000 × $25,000 = $15,000
Apportioned to Finishing department
= 800 ÷ 2,000 × $25,000 = $10,000
b) Supervision is apportioned to both departments based on numbers of Employees.
Total Supervision = $60,000
Total numbers of Employees = 10
Apportioned to Assembly department
= 4 ÷ 10 × $60,000 = $24,000
Apportioned to Finishing department
= 6 ÷ 10 × $60,000 = $36,000

(3) **Reapportionment**
Production cost centers usually use the services provided by service cost centers. Service cost centers are those that aim to provide service to other cost centers. Service cost centre do not directly involve in producing process. Finally, their cost must be re-apportioned to production cost centre.

Example 2.12

Sunshine Plc. is divided into four departments. A and B are production departments, stores and maintenance are service departments. Actual costs for the period are as follows:

		$
Allocated costs	A	25,000
	B	18,000
	stores	6,000
	administration	12,000
Apportioned costs	A	15,000
	B	10,000
	stores	2,000
	maintenance	3,000

Department A worked 8,000 labor hours and requisitioned materials worth $12,000.

Department B worked 12,000 labor hours and requisitioned materials worth $8,000.

Required:

Show the total production overheads of department A and B.

Solution:

Total production overheads of department A and B as showed in table 2.6:

Table 2.6   Total Production Overheads of Department A and B

Department	A	B	Stores	Maintenance
Allocated	25,000	18,000	6,000	12,000
Apportioned	15,000	10,000	2,000	3,000
Total All. &App.	40,000	28,000	8,000	15,000
Re-apportioned				
Maintenance	6,000	9,000	—	(15,000)
Stores	4,800	3,200	(8,000)	—
Total POH	50,800	40,200	—	—

2.3.4 Bases of Absorption

The main purpose of absorption costing is to find a way of estimating the amount of overhead that a product will absorb. Overhead absorption rate (OAR) is such an attempt at coming up with the best 「guess」 of how much overhead should be given to a product.

The formula to calculate OAR is:

OAR = Budgeted overhead/Budgeted base

Common examples of budgeted bases are machine hours or

direct labor hours (the number of hours for which a machine is in production). The choice of overhead absorption base may be made with the objective of obtaining 「accurate」 product costs. In a machine based or automated production environment, taking finishing overhead as an example, the budgeted machine hour is an appropriate basis for absorption. While in a labor intensive production environment, taking assembly overhead as an example, budgeted labor hour is the best basis for absorption.

Example 2. 13

A company has two production cost centres, manufacturing and finishing. The overheads for these two cost centers are as follows:

	Manufacturing	Finishing
Budgeted overhead	$65,000	$32,000
Labor hours	7,500	6,100
Machine hours	2,800	4,500

Management have decided that the overheads are to be absorbed based on labor hours for manufacturing overheads and to be absorbed based on machine hours for finishing overheads.

One product, the STAR, has the following details:

	Manufacturing	Finishing
Labor hours per unit	6	3
Machine hours per unit	8	2

Required:

Calculate the OAR for two cost centers respectively.

Calculate the total overhead would be included in the cost of one unit of STAR.

Solution:

OAR (Manufacturing)

　　= Budgeted overhead/ Labor hours

　　= $65,000÷7,500 = $8.7/Lhr

OAR (Finishing)

　　= Budgeted overhead/Machine hours

　　= $32,000÷4,500 = $7.1/Mhr

Manufacturing overhead = $8.7/Lhr×6hr = $52.2

Finishing overhead = $7.1/Mhr×2hr = $14.2

Total overhead = $52.2 + $14.2 = $66.4

2.3.5 The Reason of Using A Predetermined Overhead Rate

Notice that the procedure of overhead absorption described above is based on an estimated overhead rate (predetermined overhead rate), rather than an actual manufacturing overhead cost incurred by the job. The reason is that the total actual manufacturing overhead costs are usually not known to managers before the end of the year. The application of manufacturing overhead based on a predetermined overhead rate helps in computing cost of goods sold of a particular product.

2.3.6 Ledger Entries for Manufacturing Overheads

The direct costs of production are debited in the work-in-progress (WIP) account. Indirect production costs are collected in the production overheads account.

Non-production overheads are debited to accounts such as: Administration overheads account, selling overheads account, distribution overheads account and finance overheads account.

Absorbed production overheads are credited to the production overheads account.

The under-absorbed or over-absorbed overhead probably occurs if the actual overheads does not equal to absorbed overheads. The difference between them is transferred to the income statement at the end of an accounting period.

2.3.7 Under- and Over-absorption of Overheads

Since OAR is predetermined, it may lead to a result that the actual fixed costs are not properly accounted for and that a final adjustment is needed in respect of under-absorption or over-absorption of overheads.

Over-absorption:
  Overheads absorbed > Actual overheads
Under-absorption:
  Overheads absorbed < Actual overheads

Example 2.14

Budgeted fixed costs = $50,000
Budgeted output = 25,000 units
Scenario (1):
Actual fixed costs = $60,000
Actual production = 26,500 units
Scenario (2):
Actual fixed costs = $47,500
Actual production = 24,500 units

What final adjustment is needed in respect of under-absorption and over-absorption in scenario (1) and (2)?

Solution:
OAR = $50,000/25,000$ units = $2
In scenario (1), the 26,500 units produced will each absorb $2, thus, $2×26,500 = $53,000 fixed costs will be accounted for in the cost of production. However, actual fixed costs are $60,000, so the costs have been under-absorbed by $7,000 and a deduction of this amount has to be made from profits.

In scenario (2), the 24,500 units produced will each absorb $2, thus, $2×24,500 = $49,000 fixed costs will be accounted for in the cost of production.
However, actual fixed costs are $47,500, so the costs have been over-absorbed by $1,500 and $1,500 has to be added back to profits.

2.3.8 Relating Non-production Overheads to Cost Units

(1) Non-production Overheads

As we have discussed, production overhead are overhead items necessary to produce your product or service, such as the square footage necessary to house your production equipment and area. Non-production overhead will include items not directly related to production. They are the indirect costs of an organization that are not classified as manufacturing overhead, including administration overheads, selling overhead, distribution overhead, and (in some cases) research and development costs.

The cost card below shows how the total cost of a unit of product is building up from direct and indirect costs.

Cost card is a sheet of the total cost of one unit of a product.

	$
Direct materials	×
Direct labor	×
Direct expenses	×
Prime cost	×
Variable production overhead	×
Marginal production cost	×
Fixed production overheads	×

Total production cost	×
Non-production overheads	
Administration	×
Selling	×
Distribution	×
TOTAL COST	×
Profit	×
Sales price	×

(2) Relating Non-production Overheads to Cost Units

The most commonly used method of relating non-production overheads to cost units is to allocate non-production overheads on the basis of the proportion of production cost.

Example 2.15

A company manufactures Product DIY which has a production cost of $16 per unit. This company has planed production costs of $28,000 for the manufacture of Product DIY and planed non-production overheads of $7,000 associated with the production of Product X.

Required:

Calculate the total cost of product DIY if non-production overheads are related to cost units on the basis of a proportion of production costs of Product DIY.

Solution:

Non-production overhead per unit = Budgeted Non-production cost / Budgeted production cost

= $7,000 : $28,000 = 25%

Total cost of product DIY

= Production cost + Non-production overhead

= [$16 + (25% × $16)] = $20

QUESTIONS:

1. A manufacturing company uses 28,000 components at an even rate during the year. Each order placed with the supplier of the components is for 1,500 components, which is the economic order quantity. The company holds a buffer inventory of 700 components. The annual cost of holding one component in inventory is $3.

What is the total annual cost of holding inventory of the component?

A. $2,250

B. $3,300
C. $4,350
D. $4,500

2. Perpetual inventory is the counting and valuing of selected items on a rotating basis. This statement is:
A. True
B. False

3. Which of the following procedures are carried out to minimise losses from inventory?
(i) Use of standard costs for purchases.
(ii) Restricted access to stores.
(iii) Regular stocktaking.
A. (i) and (ii)
B. (ii) and (iii)
C. (ii) only
D. All of them

4. The company is trying to figure out the optimal order quantity of raw material X. Information is given below:

Annual usage	48,000 units
Purchase price	$80 per unit
Ordering costs	$120 per order
Annual holding costs	10% of the purchase price

What is the optional order quantity of material X?
A. 438
B. 800
C. 1,200
D. 3,795

5. An overhead absorption rate is used to:
A. Share out common costs over benefiting cost centres
B. Find the total overheads for a cost centre
C. Charge overheads to products
D. Control overheads

6. A manufacturing firm is very busy and overtime is being worked. The amount of overtime premium contained in direct wages would normally be classed as:
A. Part of prime cost
B. Factory overheads
C. Direct labour costs
D. Administrative overheads

7. A firm absorbs production overheads using a direct labor hour basis. Budgeted production overheads for the year just ended were $268,800 and budgeted labor hour were 48,000. The actual production overheads were $245,600 and actual laborhours were 45,000. Hence, the production overhead costs were _____.
A. Under-absorbed by $6,400
B. Under-absorbed by $16,800
C. Over-absorbed by $6,400
D. Over-absorbed by $16,800

8. The double entry for an issue of indirect production materials would be:
A. Dr Materials control account   Cr Finished goods control account
B. Dr Production overhead control account   Cr Materials control account
C. Dr Work-in-progress control account   Cr Production overhead control account
D. Dr Work-in-progress control account   Cr Materials control account

9. A factory has two production departments, X and Y, and two service departments C and D. The following information costs relates to the overhead costs in each department.

| | Manufacturing departments || Service departments ||
	X	Y	C	D
Overhead costs	$5,000	$7,500	$3,200	$4,600
Proportion of usage of services of C	50%	40%	—	10%
Proportion of usage of services of D	20%	60%	20%	—

Using the reciprocal method of apportioning service department costs, the total overhead costs allocated to department X will be:
A. $5,000
B. $7,520
C. $8,105
D. $12,195

10. The following represent transactions on the material account for a company for the month of March 2018:

	$000s
Issued to production	144
Returned to stores	5

The material inventory at 1 March 2018 was $23,000 and at 31 March 2018 was $15,000. How much material was purchased in March 2018?
A. $131,000
B. $139,000
C. $141,000
D. $159,000

Chapter 3　Absorption Costing and Marginal Costing

Learning Objectives

After the study of this chapter, you should be able to:
a) Distinguish between Absorption Costing and Marginal Costing.
b) Calculate profit by Absorption Costing and Marginal Costing.
c) Reconcile profit or losses calculated under absorption and marginal costing.
d) Describe the advantages and disadvantages of absorption and marginal costing.

3.1　Absorption Cost and Absorption Costing

3.1.1　Introduction

吸收成本(Absorption cost)又稱完全成本。完全成本是指產品成本不僅包含直接材料、直接人工、制造費用,還包括管理和銷售費用等。

吸收成本法(Absorption Costing)又稱完全成本法,是產品成本計算的一種方法,在這種方法下,產品成本包含了全部成本。

The objective of absorption costing is to include in the total cost of a product an appropriate share of the organization's total overhead. An appropriate share generally means an amount which reflects the amount of time and effort that has devoted to producing a unit or completing a job.

An organization with one production department that produces identical units will divide the total overheads among the total units produced. Absorption costing is a method for sharing overheads between different products on a fair basis.

Example 3.1

Which of the following are acceptable bases for absorbing production overheads?
(i) Direct labour hours.
(ii) Machine hours.
(iii) As a percentage of the prime cost.
(iv) Per unit.
A. Method (i) and (ii) only
B. Method (iii) and (iv) only
C. Method (i), (ii), (iii) and (iv)
D. Method (i), (ii) or (iii) only

Solution:

The correct answer is C.

All of the methods are acceptable bases for absorbing production overheads. However, the percentage of prime cost has serious limitations and the rate per unit can only be used if all cost units are identical.

3.1.2 Theoretical Reasons for Using Absorption Costing

The theoretical justification for using absorption costing is that all production overheads are incurred in the production of the organisation's output and so each unit of the product receives some benefit from these costs. Each unit of output should therefore be charged with some of the overhead costs.

3.1.3 Practical Reasons of Using Absorption Costing

The main reasons for using absorption costing are inventory valuations, pricing decisions, and establishing the profitability of different products.

(1) Inventory Valuations

Inventory in hand must be valued for two reasons:

a) For the closing inventory figure in the statement of financial position.

b) For the cost of sales figure in the income statement.

The valuation of inventory will affect profitability during a period because of the way in which the cost of sales is calculated.

The cost of goods produced + The value of opening inventories − The value of closing inventories = The cost of goods sold.

For example, closing inventories might be valued at prime cost, but in absorption costing, they would be valued at a fully absorbed factory cost. (They would not be valued at the full cost of sales, because the only costs incurred in producing goods for finished inventory are factory costs.)

(2) Pricing Decisions

Many companies attempt to fix selling prices by calculating the full cost of production or sales of each product, and then adding a margin for profit. For example, if factory cost is $8 per unit, the company might have fixed a gross profit margin at 25% on factory cost, or 20% of the sales price, in order to establish the unit sales price of $10.「Full cost plus pricing」can be particularly useful for companies which do jobbing or contract work, where each job or contract is different, so that a standard unit sales price cannot be fixed. Without using absorption costing, a full cost is difficult to ascertain.

(3) Establishing the Profitability of Different Products

This argument in favour of absorption costing is more contentious, but is worthy of mention here. If a company sells more than one product, it will be difficult to judge how profitable each individual product is, unless overhead costs are shared on a fair basis and charged to the cost of sales of each product.

Example 3.2

Which of the following is absorption costing concerned with?
A. Direct materials   B. Direct labor
C. Fixed costs   D. Variable and fixed costs

Solution:

The correct answer is D.

Absorption costing is concerned with including in the total cost of a product an appropriate share of overhead, or indirect cost. Overheads can be fixed or variable costs, therefore option D is correct. Option A and option B are incorrect because they relate to direct costs. Option C is incorrect because it does not take account of variable overheads.

3.1.4 International Accounting Standard 2

Absorption costing is recommended in financial accounting by international accounting standard 2 (IAS 2) Inventories. IAS 2 deals with financial accounting systems. The cost accountant is (in theory) free to value inventories by whatever method seems best, but how companies integrate their financial accounting and cost accounting systems into a single system of accounting records, the valuation of closing inventories will be determined by IAS 2.

IAS 2 states that costs of all inventories should comprise those costs which have been incurred in the normal course of business in bringing the inventories to their ⌈present location and condition⌋. These costs incurred will include all related production overheads, even though these overheads may accrue on a time basis. In other words, in financial accounting, closing inventories should be valued at full factory cost, and it may therefore be convenient and appropriate to value inventories by the same method in the cost accounting system.

3.1.5 Absorption Costing Stages

The three stages of absorption costing are:
a) 1 Allocation.
b) 3 Absorption.

c) 2 Apportionment.

3.2 Marginal Cost and Marginal Costing

3.2.1 Introduction

邊際成本(Marginal Cost)又稱變動成本,是指隨著業務量變化而變化的成本。

邊際成本法(Marginal Costing)又稱變動成本法,是指計算產品成本時,產品成本僅包含隨著產量變化而變化的那部分成本,固定成本不作為產品成本而作為期間費用處理。

Marginal costing is an alternative method of costing to absorption costing. In marginal costing, only variable costs are charged as a cost of sales and a contribution is calculated (sales revenue minus variable cost of sales). Closing inventories of work in progress or finished goods are valued at marginal (variable) production cost. Fixed costs are treated as a period cost, and are charged in full to the profit and loss account of the accounting period in which they are incurred.

The marginal production cost per unit of an item usually consists of the following.
a) Direct materials.
b) Variable production overheads.
c) Direct labour.

Direct labour costs might be excluded from marginal costs when the work force is a given number of employees on a fixed wage or salary. Even so, it is not uncommon for direct labour to be treated as a variable cost, even when employees are paid a basic wage for a fixed working week. If in doubt, you should treat direct labour as a variable cost unless given clear indications to the contrary. Direct labour is often a step cost, with sufficiently short steps to make labour costs act in a variable fashion. The marginal cost of sales usually consists of the marginal cost of production adjusted for inventory movements plus the variable selling costs, which would include items such as sales commission, and possibly some variable distribution costs.

3.2.2 The Principles of Marginal Costing

The principles of marginal costing are as follows.
a) Period fixed costs are the same, for any volume of sales and production (provided that the level of activity is within the「relevant range」). Therefore, by selling an extra item of product or service the following will happen.
Revenue will increase by the sales value of the item sold.
Costs will increase by the variable cost per unit.
Profit will increase by the amount of contribution earned from the extra item.
b) Similarly, if the volume of sales falls by one item, the

profit will fall by the amount of contribution earned from the item.

c) Profit measurement should therefore be based on an analysis of total contribution. Since fixed costs relate to a period of time, and do not change with increases or decreases in sales volume, it is misleading to charge units of sale with a share of fixed costs. Absorption costing is therefore misleading, and it is more appropriate to deduct fixed costs from total contribution for the period to derive a profit figure.

d) When a unit of product is made, the extra costs incurred in its manufacture are the variable production costs. Fixed costs are unaffected, and no extra fixed costs are incurred when output is increased. It is therefore argued that the valuation of closing inventories should be at variable production cost [direct materials, direct labour, direct expenses (if any) and variable production overhead] because these are the only costs properly attributable to the product.

3.2.3 Profit or Contribution Information

邊際貢獻（Contribution）：變動成本法下，銷售收入-變動成本=邊際貢獻

The main advantage of contribution information (rather than profit information) is that it allows an easy calculation of profit if sales increase or decrease from a certain level. By comparing total contribution with fixed overheads, it is possible to determine whether profits or losses will be made at certain sales levels. Profit information, on the other hand, does not lend itself to easy manipulation but note how easy it was to calculate profits using contribution information in the question entitled marginal costing principles. Contribution information is more useful for decision making than profit information.

3.3 Absorption Costing, Marginal Costing and the Calculation of Profit

3.3.1 Introduction

In marginal costing, fixed production costs are treated as period costs and are written off as they are incurred. In absorption costing, fixed production costs are absorbed into the cost of units and are carried forward in inventory to be charged against sales for the next period. Inventory values using absorption costing are therefore greater than those calculated using marginal costing.

Marginal costing as a cost accounting system is significantly different from absorption costing, please look at table 3.1

as following. It is an alternative method of accounting for costs and profit, which rejects the principles of absorbing fixed overheads into unit costs.

Table 3.1　　Contrasting Marginal Costing and Absorption Costing

Marginal Costing	Absorption Costing
Closing inventories are valued at marginal production cost	Closing inventories are valued at full production cost
Fixed costs are period costs	Fixed costs are absorbed into unit costs
Cost of sales does not include a share of fixed overheads	Cost of sales does include a share of fixed overheads (see note below)

Note: The share of fixed overheads included in cost of sales is from the previous period (in opening inventory values). Some of the fixed overheads from the current period will be excluded by being carried forward in closing inventory values.

In marginal costing, it is necessary to identify the following.
a) Variable costs.
b) Fixed costs.
c) Contribution.

In absorption costing (sometimes known as full costing), it is not necessary to distinguish variable costs from fixed costs.

3.3.2　The Calculating Profit by Absorption Costing

Example 3.3

完全成本法下利潤的計算：銷售收入 – 銷售成本 = 毛利

注意：固定成本已經分配到銷售成本當中。

Cost and selling price details for product Z are as follows.

	$ per unit
Direct materials	6.00
Direct labour	7.50
Variable overhead	2.50
Fixed overhead absorption rate	5.00
	21.00
Profit	9.00
Selling price	30.00

Budgeted production for the month was 5,000 units although the company managed to produce 5,800 units, selling 5,200 of them and incurring fixed overhead costs of $27,400.

What is the absorption costing profit for the month?
A. $45,200　　C. $46,800
B. $45,400　　D. $48,400

Solution:
The correct answer is D.

	$	$
Sales(5,200 at $30)		156,000
Materials (5,200 at $6)	31,200	
Labor (5,200 at $7.50)	39,000	
Variable overhead (5,200 at $2.50)	13,000	
Total variable cost		(83,200)
Fixed overhead (5,200 at $5)		(26,000)
Over-absorbed overhead (W)		1,600
Absorption costing profit		48,400

Working:

	$
Overhead absorbed (5,800 at $5)	29,000
Overhead incurred	27,400
Over-absorbed overhead	1,600

3.3.3 The Calculating Profit by Marginal Costing

變動成本法下利潤計算:銷售收入-變動成本=邊際貢獻

注意:固定成本作為期間費用處理。

Example 3.4

The following example will be used to lead you through the various steps in calculating Marginal costing profits.

Big Woof manufactures a single product, the Bark, details of which are as follows:

Per unit	$
Selling price	180.00
Direct materials	40.00
Direct labor	16.00
Variable overheads	10.00

Annual fixed production overheads are budgeted to be $1.6 million and Big Woof expects to produce 1,280,000 units of the Bark each year. Overheads are absorbed on a per unit basis. Actual overheads are $1.6 million for the year. Budgeted fixed selling costs are $320,000 per quarter. Actual sales and production units for the first quarter of 20X8 are given below.

January - March
Sales   240,000
Production   280,000

There is no opening inventory at the beginning of January. Prepare income statements for the quarter, using Marginal costing.

Solution:
Step 1: Overhead absorption rate = Budgeted units/Budge-

ted fixed overheads

Also be careful with your calculations. You are dealing with a three-month period but the figures in the question are for a whole year. You will have to convert these to quarterly figures.

Budgeted overheads (quarterly) = $1,600,000÷4
= $400,000
Budgeted production (quarterly) = 1,280,000÷4
= 320,000 units
Overhead absorption rate per unit = $400,000÷320,000
= $1.25 per unit

Step 2: Calculate total cost per unit
Total cost per unit (absorption costing)
= Variable cost + Fixed production cost
= (40 + 16 + 10) + 1.25
= $67.25
Total cost per unit (marginal costing)
= Variable cost per unit
= $66

Step 3: Calculate closing inventory in units
Closing inventory
= Opening inventory + Production − Sales
= 0 + 280,000 − 240,000
= 40,000 units

Step 4: Calculate under/over absorption of overheads
This is based on the difference between actual production and budgeted production.
Actual production = 280,000 units
Budgeted production = 320,000 units (see step 1 above)
Under-Production = 40,000 units
As Big Woof produced 40,000 fewer units than expected, there will be an under absorption of overheads of 40,000 × $1.25 (see step 1 above) = $50,000. This will be added to production costs in the income statement.

Step 5: Prepare income statements

	$'000	$'000
Sales (240,000 × $180)		43,200
Less Cost of Sales		

```
                Opening inventory                              0
                Add Production cost (280,000 × $66)      18,480
                Less Closing inventory (40,000 × $66)    (2,640)
                                                                (15,840)
                Contribution                                     27,360
                Less fixed O/H
                Fixed production O/H                       400
                Fixed selling O/H                          320
                                                                  (720)
                Net profit                                       26,640
```

3.4 Reconciling Profits

3.4.1 Introduction

随著期初存貨量和期末存貨量的變化,完全成本法和變動成本法計算出來的利潤額可能會不一致。利潤差額可以用以下公式計算:

利潤差額＝存貨增減額×單位產品費用分配率

Reported profit figures using marginal costing or absorption costing will differ if there is any change in the level of inventories in the period. If production is equal to sales, there will be no difference in calculating profits with different costing methods.

The difference in profits reported under the two costing systems is due to the different inventory valuation methods used.

If inventory levels increase between the beginning and end of a period, absorption costing will report the higher profit. This is because some of the fixed production overhead incurred during the period will be carried forward in closing inventory (which reduces cost of sales) to be set against sales revenue in the following period instead of being written off in full against profit in the period concerned.

If inventory levels decrease, absorption costing will report the lower profit because as well as the fixed overhead incurred, fixed production overhead which had been carried forward in opening inventory is released and is also included in cost of sales.

A quick way to establish the difference in profits without going through the whole process of drawing up the income statements is as follows:

Difference in profits = Change in inventory level × Overhead absorption rate per unit

If inventory levels have gone up (closing inventory > opening inventory), then absorption costing profit will be greater than marginal costing profit.

If inventory levels have gone down (closing inventory < opening inventory), then absorption costing profit will be less than marginal costing profit.

3.4.2 Reconciling Profits

Example 3.5

Last month a manufacturing company's profit was $2,000, calculated using absorption costing principles. If marginal costing principle has been used, a loss of $3,000 would have occurred. The company's fixed production cost is $2 per unit. Sales last month were 10,000 units.
What was last month's production (in units)?
A. 7,500　　B. 9,500　　C. 10,500　　D. 12,500

Solution:
The correct answer is D.
Any difference between marginal and absorption costing profit is due to changes in inventory.

	$
Absorption costing profit	2,000
Marginal costing loss	(3,000)
Difference	5,000

Change in inventory = Difference in profit/Fixed product cost per unit = $5,000/$2 = 2,500 units

Marginal costing loss is lower than absorption costing profit, therefore inventory has gone up, production was greater than sales by 2,500 units.

Production = 10,000 units (sales) + 2,500 units
　　　　　= 12,500 units

QUESTIONS:

1. Which of the following are true of marginal costing?
(i) The marginal cost of a product includes an allowance for fixed production costs.
(ii) The marginal cost of a product represents the additional cost of producing an extra unit.
(iii) If the inventory increases over a year, the profits under absorption costing will be lower than with marginal costing.
A. (i) only
B. (ii) only
C. (ii) and (iii) only
D. (i), (ii) and (iii)

2. A company produces and sells a single product whose variable cost is $6 per unit.
Fixed costs have been absorbed over the normal level of activity of 200,000 units and have been calculated as $2 per unit.
The current selling price is $10 per unit.
How much profit is made under marginal costing if the company sells 250,000 units?
A. $500,000
B. $600,000
C. $900,000
D. $1,000,000

3. Exp has compiled the following standard cost card for its main product.

	$
Production costs	
Fixed	33.00
Variable	45.10
Selling costs	
Fixed	64.00
Variable	7.20
Profit	14.70
Selling price	164.00

Under an absorption costing system, closing inventory would be valued at:
A. $52.30   B. $78.10
C. $97.00   D. $149.30

4. Consider the following figure 3.1 for total costs and total revenue:

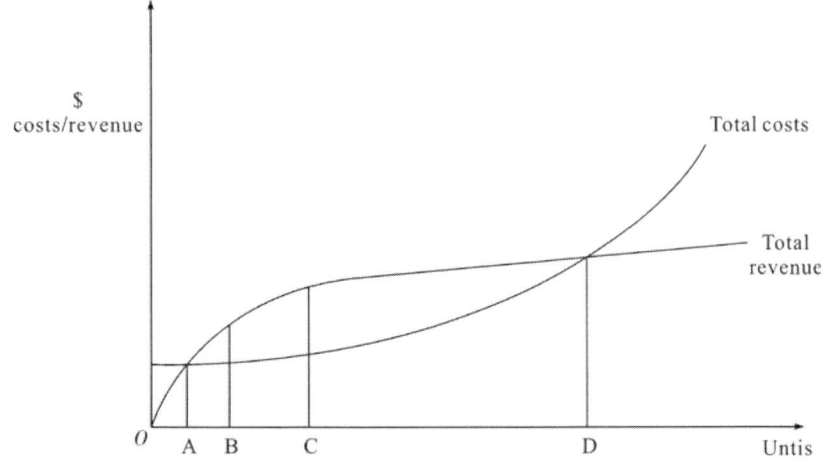

Figure 3.1 Total Costs and Total Revenue

At which point on the above graph is it most likely that profits will be maximised?
A. A   B. B   C. C   D. D

5. Assume that at the end of the first month unit variable costs and fixed costs and selling price for the month were in line with the budget and any inventory was valued at the same unit cost as in the above budget.
However, if production was actually 700 and sales 600, what would be the reported profit using absorption costing?
A. $9,000
B. $12,000
C. $14,000
D. $15,000

6. Which of these statements are true of marginal costing?
(i) The contribution per unit will be constant if the sales volume increases.
(ii) There is no under-absorption or over-absorption of overheads.
(iii) Marginal costing does not provide useful information for decision making.
A. (i) and (ii) only
B. (ii) and (iii) only
C. (ii) only
D. (i), (ii) and (iii)

7. Last month a manufacturing company's profit was $2,000, calculated using absorption costing principles. If marginal costing principles had been used, a loss of $3,000 would have occurred. The company's fixed production cost is $2 per unit. Sales last month were 10,000 units.
What was last month's production (in units)?
A. 7,500
B. 9,500
C. 10,500
D. 12,500

8. A company manufactures and sells a single product. For this month the budgeted fixed production overheads are $48,000, budgeted production is 12,000 units and budgeted sales are 11,720 units.
The company currently uses absorption costing.
If the company used marginal costing principles instead of absorption costing for this month, what would be the effect on the budgeted profit?
A. $1,120 higher
B. $1,120 lower
C. $3,920 higher
D. $3,920 lower

9. When opening inventory was 8,500 litres and closing inventory was 6,750 litres, a firm had a profit of $62,100 using marginal costing.
Assuming that the fixed overhead absorption rate was $3 per litre, what would be the profit

using absorption costing?

A. $41,850
B. $56,850
C. $67,350
D. $82,350

10. In a given period, the production level of an item exactly matches the level of sales. The profit reported will be identical whether marginal or absorption costing is used.
This statement is:
A. True
B. False

Chapter 4　Cost Accounting Methods

Learning Objectives

After the study of this chapter, you should be able to:
a) Distinguish between job costing and process costing.
b) Calculate material, labour and overhead in a job order costing.
c) Calculate material, labour and overhead in a process costing.
d) Calculate unit costs for a service company.

4.1　Job Costing

4.1.1　Introduction

　　Job Costing 和 Job Order Costing 直譯為工作成本法。在中文中, Job Costing 和 Batch Costing 通常翻譯為分批法。

Job costing is a costing method applied where work is undertaken to customers' special requirements and each order is of comparatively short duration. A job is a cost unit which consists of a single order or contract. The work relating to a job moves through processes and operations as a continuously identifiable unit. Job costing is most commonly applied within a factory or workshop, but may also be applied to property repairs and internal capital expenditure.

Example 4.1

Which of the following costing methods is most likely to be used by a company involved in the manufacture of furniture?
A. Batch costing
B. Service costing
C. Job costing
D. Process costing

Solution:
The correct answer is C.
Job costing is a costing method applied where work is undertaken to customers' special requirements and each order is of comparatively short duration.

4.1.2 Procedure for the Performance of Jobs

The normal procedure in jobbing concerns involves:

a) The prospective customer approaches the supplier and indicates the requirements of the job.

b) A representative sees the prospective customer and agrees with him the precise details of the items to be supplied, for example, the quantity, quality, size and the colour of the goods, the date of delivery and any special requirements.

c) The estimating department of the organisation then prepares an estimate for the job. This will be based on the cost of materials to be used, the labour expense expected, the cost overheads, the cost of any additional equipment needed specially for the job, and finally the supplier's profit margin. The total of these items will represent the quoted selling price.

d) If the estimate is accepted the job can be scheduled. All materials, labour and equipment required will be ⌈booked⌋ for the job. In an efficient organisation, the start of the job will be timed to ensure that it will not be loaded too early while it is ready for delivery to the customers, otherwise storage space will have to be found for the product until the date required by the customer the customer.

4.1.3 Job Cost Sheets/Cards

Costs for each job are collected on a job cost sheet or job card. With other methods of costing, it is usual to produce for inventory; This means that management must decide in advance how many units of each type, size, colour, quality and so on which will be produced during the coming year, regardless of which kind of customers will eventually buy the product. In job costing, because production is usually carried out in accordance with the special requirements of each customer, it is usual for each job to differ in one or more respects from another.

Therefore, a separate record must be maintained to show the details of individual jobs. Such records are often known as job cost sheets or job cost cards. An example is shown on the table 4.1. Either the detail of relatively small jobs or a summary of direct materials, direct labour and so on for larger jobs will be shown on a job cost sheet.

Table 4.1　　　　Job Cost Sheets

Items	$
Direct Material Costs ($/u×Qty used)	×
Direct Labour Costs ($/hr×Hrs spent)	×
Direct Expenses	×
Prime Costs per job	×
Production Overheads (OAR×Actual base)	×
Total Production Costs per job	×
Non-production Overheads (OAR×Actual base)	×
Total Costs per job	××

4.1.4 Job Cost Information

Information for the direct and indirect costs will be gathered as follows.

(1) Direct Material Cost
a) The estimated cost will be calculated by valuing all items in the bill of materials. Materials that have to be specially purchased for the job in question will need to be priced by the purchasing department.
b) The actual cost of materials used will be calculated by valuing materials issues notes from storage job or invoices for materials specifically purchased. All documentation should indicate the job number to which it relates.

(2) Direct Labour Cost
a) The estimated labour time required will be calculated from past experience of similar types of work or work study engineers may prepare estimates following detailed specifications. Effected factors in labour rates like Increases, overtime and bonuses should be taken into account.
b) The actual labour hours will be available from either time sheets or job tickets/cards, using job numbers where appropriate to indicate the time spent on each job. The actual labour cost will be calculated using the hours information and current labour rates (plus bonuses, overtime payments and so on).

(3) Direct Expenses
a) The estimated cost of any expenses likely to be incurred can be obtained from a supplier.
b) The details of actual direct expenses incurred can be taken from invoices.

(4) Production Overheads
a) The estimated production overheads to be included in

the job cost will be calculated from overhead absorption rates in operation and the estimate of the basis of the absorption rate (for example, direct labour hours). This assumes the job estimate include overheads (in a competitive environment manager may feel that if overheads incurred in respective of the job is taken on or not, the minimum estimated quotation price should be based on variable costs only).

b) The actual production overhead to be included in the job cost will be calculated from the overhead absorption rate and the actual results (such as labour hours coded to the job in question). Inaccurate overhead absorption rates can seriously harm an organisation; If jobs are over priced, customers will go elsewhere and if jobs are under priced, revenue will fail to cover costs.

(5) Administration, Selling and Distribution Overheads

The organisation may absorb non-production overheads using any methods (for example, percentage on full production cost or estimates of costs and actual costs included in job cost.

4.1.5 Rectification Costs

If the finished output is found to be sub-standard, it may be possible to rectify the fault. The sub-standard output will then be returned to the department or cost centre where the fault arose.

Rectification costs can be treated in two ways.

a) If rectification work is not a frequent occurrence, but arises on occasions with specific jobs to which it can be traced directly, then the rectification costs should be charged as a direct cost to the jobs concerned.

b) If rectification is regarded as a normal part of the work carried out generally in the department, then the rectification costs should be treated as production overheads. This means that they would be included in the total of production overheads for the department and absorbed into the cost of all jobs for the period, using the overhead absorption rate.

4.1.6 Work in Progress

Work in Progress 直譯為在產品，在產品典型估價方法是約當產量法。

At the year end, the value of work in progress is simply the sum of the costs incurred on incomplete jobs (provided that the costs are lower than the net realisable value of the customer order).

Example 4.2

A firm makes special assemblies to customers' orders and uses job costing.

The data for a period are shown in table 4.2 as following:

Table 4.2   **The Date for A Period**

	Job number		
	AA10($)	BB15($)	CC20($)
Opening WIP	26,800	42,790	0
Material added in period	17,275	0	18,500
Labour for period	14,500	3,500	24,600

The budgeted overheads for the period were $126,000.

Job number BB15 was completed on the last day of the period.

a) What overhead should be added to job number CC20 for the period?

A. $65,157   B. $69,290
C. $72,761   D. $126,000

b) What was the approximate value of closing work-in-progress at the end of the period?

A. $58,575   B. $101,675
C. $217,323   D. $227,675

Solution:

a) The correct answer is C.

The most logical basis for absorbing the overhead job costs is to use a percentage of direct labour cost.

Overhead = 24,600 ÷ (14,500 + 3,500 + 24,600) × 126,000
= 72,761

b) The correct answer is C, the analysis is shown in table 4.3 as following:

Table 4.3   **The Analysis**

Job number	WIP($)
AA10 (26,800 + 17,275 + 14,500) + (14,500 ÷ 42,600 × 126,000)	101,462
CC20 (18,500 + 24,600 + 72,761)	115,861
Σ	217,323

4.1.7 Pricing the Job

典型的工作（產品）定價方法是成本加成法。

價格＝成本＋利潤

The usual method of fixing prices in a jobbing concern is cost plus pricing. Cost plus pricing means that a desired profit margin is added to total costs to arrive at the selling price. The estimated profit will depend on the particular circumstance of the job and organisation in question. In competitive situations, the profit may be small but if the organi-

Example 4.3

A company calculates the prices of jobs by adding overheads to the prime cost and adding 30% to total costs as a mark up. Job number Y256 was sold for $1,690 and incurred overheads of $694. What was the prime cost of the job?

A. $489   B. $606   C. $996   D. $1,300

Solution:

The correct answer is B.

	$
Selling price of job	1,690
Less profit margin (30/130)	390
Total cost of job	1,300
Less overhead	694
Prime cost	606

4.1.8 Job Costing for Internal Services

It is possible to use a job costing system to control the costs of an internal service department, such as the maintenance department or the printing department. If a job costing system is used, it is possible to charge the user departments for the cost of specific jobs carried out, rather than apportioning the total costs of these service departments to the user departments using an arbitrarily determined apportionment basis.

An internal job costing system for service departments will have the following advantages, please look at table 4.4 as following:

Table 4.4 The Advantages of A Internal Job Costing System for Service Departments

Advantages	Comment
Realistic apportionment	The identification of expenses with jobs and the subsequent charging of these to the department(s) responsible means that costs are borne by those who incurred them
Increased responsibility and awareness	User departments will be aware that they are charged for the specific services used and may be more careful to use the facility more efficiently. They will also appreciate the true cost of the facilities that they are using and can take decisions accordingly

Table 4. 4 (continued)

Advantages	Comment
Control of service department costs	The service department may be restricted to charging a standard cost to user departments for specific jobs carried out or time spent. It will then be possible to measure the efficiency or inefficiency of the service department by recording the difference between the standard charges and the actual expenditure
Planning information	This information will ease the planning process, as the purpose and cost of service department expenditure can be separately identified

Example 4. 4

A firm uses job costing and recovers overheads on direct labour. Three jobs were worked on during a period, the details of which are showed in table 4. 5 as follows:

Table 4. 5   **The Details of Three Jobs**

	Job number		
	Job 1 ($)	Job 2 ($)	Job 3 ($)
Opening work in progress	8,500	0	46,000
Material in period	17,150	29,025	0
Labour for period	12,500	23,000	4,500

The overheads for the period were exactly as budgeted, $140,000.
Job 1 and Job 2 were the only incomplete jobs.
What was the value of closing work in progress?
A. $81,900   B. $90,175
C. $140,675   D. $214,425

Solution:
The correct answer is D.
Total labour cost = $12,500 + $23,000 + $4,500
                  = $40,000
Overhead absorption rate = 40,000 ÷ 140,000 × 100%
                         = 350% of direct labour cost
The analysis is showed in table 4. 6 as following:

Table 4. 6   **The Analysis of the Example**

	Job number		
	Job 1 ($)	Job 2 ($)	Total ($)
Costs given in question	38,150	52,025	90,175
Overhead absorbed	43,750 (12,500×350%)	80,500 (23,000×350%)	124,250
Σ	81,900	132,525	214,425

4.2 Batch Costing

Batch Costing 直譯為分批成本法。在中文中, Job Costing 和 Batch Costing 通常翻譯為分批法。

Batch costing is used when a business is manufacturing for a specific order or stock purposes, where the production is in batch. A batch is a group of similar articles which maintains its identity during one or more stages of production and is treated as a cost unit. Batch costing is similar to job costing in many respects when a batch is produced for a specific order.

In general, the procedures for costing batches are very similar to those for costing jobs.

a) The batch is treated as a job during production and the costs are collected in the manner already described in this chapter.

b) Once the batch has been completed, the cost per unit can be calculated as the total batch cost divided into the number of units in the batch.

Example 4.5

The following items may be used in costing batches.
(i) Actual material cost.
(ii) Actual manufacturing overheads.
(iii) Absorbed manufacturing overheads.
(iv) Actual labour costs.
Which of the above are contained in a typical batch cost?
A. (i), (ii) and (iv) only   B. (i) and (iv) only
C. (i), (iii) and (iv) only   D. (i), (ii), (iii) and (iv)

Solution:
The correct answer is C.
The actual material and labour costs for a batch [(i) and (iv)] can be determined from the material and labour recording system. Actual manufacturing overheads cannot be determined for a specific batch because of the need for allocation and apportionment of each item of overhead expenditure, and the subsequent calculation of a predetermined overhead absorption rate. Therefore item (ii) is incorrect and item (iii) is correct.

4.3 Process Costing

4.3.1 The Basics of Process Costing

Process Costing 直譯為分步成本法。

(1) Introduction to Process Costing

Process costing is a costing method with the continuous nature of production processes, it is usually used in situations where identifying separate units of production or jobs is impossible.

It is common to identify process costing with continuous production such as the following.
a) Oil refining.
b) Foods and drinks.
c) Paper.
d) Chemicals.

Process costing may also be associated with the continuous production of large volumes of low-cost items, such as cans or tins.

(2) Features of Process Costing

a) The output of one process becomes the input to the next until the finished product is made in the final process.
b) The continuous nature of production in many processes means that there will usually be closing.
c) Work in progress which must be valued. In process costing it is not possible to build up cost records of the cost per unit of output or the cost per unit of closing inventory because production in progress is an indistinguishable homogeneous mass.
d) There is often a loss in process due to spoilage, wastage, evaporation and so on.
e) Output from production may be a single product, but there may also be a by-product (or byproducts) and/or joint products.

The aim of this chapter is to describe how cost accountants keep a set of accounts to record the costs of production in a processing industry. The aim of the set of accounts is to derive a cost, or valuation, for output and closing inventory.

(3) Cost Flows in Process Costing

Cost accumulation is simpler in a process costing system than in a job-order costing system. In a process costing system, instead of having to trace costs to hundreds of different jobs, costs are traced to only a few processing departments.

分步成本法中的成本流動：

首先,每一個步驟均設立一個在產品帳戶;

Chapter 4 Cost Accounting Methods 63

其次,前一個步驟的完工產品轉移到下一步驟的在產品帳戶當中;

最後,材料、人工和費用都在每一步驟增加。

A T-account model of materials, labor, and overhead cost flows in a process costing system is shown in figure 4.1. Several key points should be noted from this exhibit.

First, note that a separate Work in process account is maintained for each processing department.

Second, note that the completed production of the first processing department (Department A in the figure) is transferred to the Work in Process account of the second processing department (Department B). After further work in Department B, the completed units are then transferred to Finished Goods. (In figure 4.1, we show only two processing departments, but a company can have many processing departments.)

Finally, note that materials, labor, and overhead costs can be added in any processing department, not just the first. Costs in Department B's Work in Process account consist of the materials, labor, and overhead costs incurred in Department B plus the costs attached to partially completed units transferred in from Department A (called transferred-in costs).

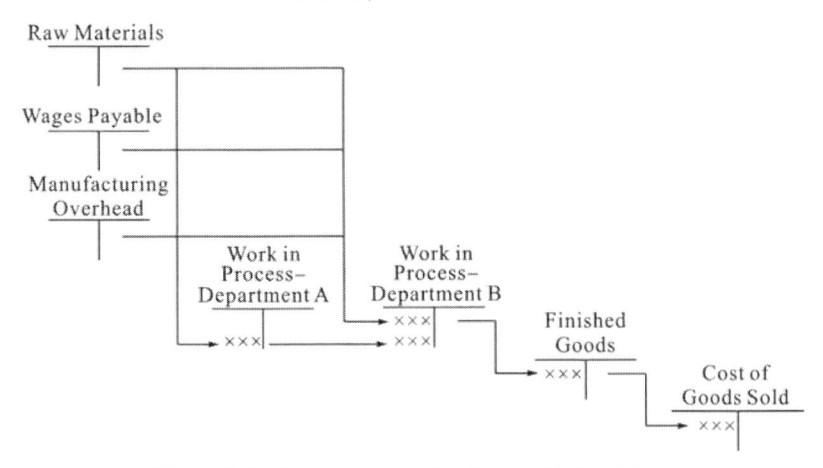

Figure 4.1 T-account Model of Process Costing Flows

分步成本法處置框架:

第一步,決定產出和損失;

第二步,計算產出、損失和在產品的單位成本;

第三步,計算產出、損失和在產品的總成本;

(4) Framework for Dealing with Process Costing

Process costing is centered around four key steps. The exact work done at each step will depend on whether there are normal losses, scrap, opening and closing work in progress.

Step 1 Determine output and losses.

This step involves the following:

a) Determining expected output.

第四步，結清帳戶。

b) Calculating normal loss and abnormal loss and gain.
c) Calculating equivalent units if there is closing or opening work in progress.
Step 2 Calculate cost per unit of output, losses and WIP.
This step involves calculating cost per unit or cost per equivalent unit.
Step 3 Calculate total costs of output, losses and WIP.
In some examples this will be straightforward; However in cases where there is closing and/or opening work-in-progress a statement of evaluation will have to be prepared.
Step 4 Complete accounts.
This step involves the following.
a) Completing the process account.
b) Writing up the other accounts required by the question.

4.3.2 Losses in Process Costing

分步成本法中的損失分為正常損失和非正常損失。

(1) Determine Output and Losses

Losses may occur in process. If a certain level of loss is expected, this is known as normal loss. If losses are greater than expected, the extra loss is abnormal loss. If losses are less than expected, the difference is known as abnormal gain.

Since normal loss is not counted as a cost, the cost of producing these units is borne by the 「good」 units of output. Abnormal loss and gain units are valued at the same unit rate as 「good」 units. Therefore, abnormal events won't affect the cost of good production. Their costs are analysed separately in an abnormal loss or abnormal gain account.

Example 4.6

A chemical process has a normal wastage of 10% of input. In a period, 2,500 kgs of material were input and there was an abnormal loss of 75 kgs.
What quantity of good production was achieved?
A. 2,175 kgs　　　B. 2,250 kgs
C. 2,325 kgs　　　D. 2,425 kgs

Solution:
The correct answer is A.
Good production = Input - Normal loss - Abnormal loss
　　　　　　　　= 2,500 - (2,500 × 10%) - 75
　　　　　　　　= 2,500 - 250 - 75
　　　　　　　　= 2,175 kgs

帶殘值的損失。

(2) Losses with Scrap Value

Scrap is「Discarded material having some value」. Loss or spoilage may have scrap value. The scrap value of normal loss is usually deducted from the cost of materials. The scrap value of abnormal loss (or abnormal gain) is usually set off against its cost, in an abnormal loss (abnormal gain) account.

Example 4.7

A company operates a continuous process into which 3,000 units of material costing $9,000 was input in a period. Conversion costs for this period were $11,970 and losses, which have a scrap value of $1.50, are expected at a rate of 10% of input. There were no opening or closing inventories and output for the period was 2,900 units. What was the output valuation?

Solution:

	$
Material	9,000
Conversion costs	11,970
Less: scrap value of normal loss (300× $1.50)	(450)
Cost of process	20,520

Expected output = 3,000 − (10%×3,000)
　　　　　　　= 3,000 − 300 = 2,700 units
Costs per unit = (Input costs−Scrap value of normal loss)/
Expected output = $20,520÷2,700 = $7.60
Value of output = 2,900× $7.60 = $22,040

帶處置成本的損失。

(3) Losses with A Disposal Cost

Loss or spoilage may have disposal cost.
The basic calculations required in such circumstances are as follows:

a) Increase the process costs by the cost of disposing of the units of normal loss and use the resulting cost per unit to value good output and abnormal loss/gain.

b) The normal loss is given no value in the process account.

c) Include the disposal costs of normal loss on the debit side of the process account.

d) Include the disposal costs of abnormal loss in the abnormal loss account and hence in the transfer of the cost of abnormal loss to the income statement.

Example 4.8

Suppose that input to a process was 1,000 units at a cost of $4,500. Normal loss is 10% and there are no opening and closing inventories. Actual output was 860 units and loss units had to be disposed of at a cost of $0.90 per unit. What is the cost per unit?

A. $5.10   B. $6   C. $5.5   D. $4.5

Solution:
The correct answer is A.
Normal loss = 10%×1,000 = 100 units
Abnormal loss = 900 − 860 = 40 units
Cost per unit = ($4,500+100×$0.90)÷900
              = $5.10

4.3.3 Valuing Closing Work in Progress

When units are partly completed at the end of a period (and hence there is closing work in progress), it is necessary to calculate the equivalent units of production in order to determine the cost of a completed unit. Equivalent units are notional whole units which represent incomplete work, and which are used to apportion costs between work in process and completed output.

在產品的估價。在產品的典型估計方法是約當產量法。

Example 4.9

Ally Co has the following information available on Process 9, please look at table 4.7.

Table 4.7   Process 9 Account

	$			$
Input	10,000kg	59,150	Finished goods	8,000kg	52,000
			Closing WIP	2,000kg	7,150
Σ		59,150	Σ		59,150

How many equivalent units were there for Closing WIP?
A. 1,000   B. 1,100
C. 2,000   D. 8,000

Solution:
The correct answer is B.
This question requires you to work backwards. You can calculate the cost per unit using the finished goods figures.
Cost per unit = Number of kg/Cost of finished goods = 52,000÷8,000 = $6.50
If 2,000kg (Closing WIP figure) were fully complete total cost would be 2,000 × $6.50 = $13,000
Actual cost of Closing WIP = $7,150

Degree of completion = 7,150 ÷ 13,000 = 55%
Therefore equivalent units = 55% × 2,000 = 1,100kg

In many industries, materials, labour and overheads may be added at different rates during the course of production.

a) Output from a previous process (for example, the output from process 1 to process 2) may be introduced into the subsequent process all at once, so that closing inventory is 100% complete in respect of these materials.

b) Further materials may be added gradually during the process, so that closing inventory is only partially complete in respect of these added materials.

c) Labour and overhead may be 「added」 at yet another different rate. When production overhead is absorbed on a labour hour basis, however, we should expect the degree of completion on overhead to be the same as the degree of completion on labour.

When this situation occurs, equivalent units, and a cost per equivalent unit should be calculated separately for each type of material, and also for conversion costs.

Example 4.10

Ashley Co operates a process costing system. The following details are available for Process 2.

Materials input at beginning of process is 12,000 kg, costing $18,000. Labour and overheads added is $28,000. 10,000kg were completed and transferred to the finished goods account. The remaining units were 60% complete with regard to labour and overheads. There were no losses in the period.

What is the value of Closing WIP in the process account?

A. $4,800   B. $6,000
C. $7,667   D. $8,000

Solution:

The correct answer is B, the analysis is shown in table 4.8 and table 4.9.

Table 4.8   **Statement of Equivalent Units**

	Material			Labour & Overhead		
	Units completion	Degree of units	Equivalent	Units completion	Degree of units	Equivalent
Finished goods	10,000	100%	10,000	10,000	100%	10,000
Closing WIP	2,000		2,000	2,000	60%	1,200
	12,000		12,000	12,000		11,200

Table 4.9　　　　　　　　　　**Costs Per Equivalent Unit**

	Material	Labour & Overhead
Total cost	18,000	28,000
Equivalent units	12,000	11,200
Cost per unit	$1.50	$2.50

Total cost per unit = $4.00
Value of Closing WIP
= ($1.50 × 2,000) + ($2.50 × 1,200) = $6,000

4.3.4 Valuing Opening Work in Progress

期初在產品估價。期初在產品計價方法可採用先進先出法、後進先出法和平均法等。

Account should be taken of opening work in progress using either the FIFO method or the weighted average cost method. Opening work in progress is partly completed at the beginning of a period and is valued at the cost incurred to date. The FIFO method of valuation deals with production on a first in, first out basis. The assumption is that the first units completed in any period are the units of opening inventory that were held at the beginning of the period.

Example 4.11

Walter Co uses the FIFO method of process costing. At the end of a four week period, the following information was available for process P.

Opening WIP　　2,000 units (60% complete) costing $3,000 to date
Closing WIP　　1,500 units (40% complete)
Transferred to next process　　7,000 units

How many units were started and completed during the period?

A. 5,000 units　　　　　C. 8,400 units
B. 7,000 units　　　　　D. 9,000 units

Solution:
The correct answer is A.
As we are dealing with the FIFO method, Opening WIP must be completed first.
Total output　　　　　　　　　　　　　　　　7,000 units
Less Opening WIP (completed first)　　　　(2,000 units)
Units started and completed during the period
　　　　　　　　　　　　　　　　　　　　　5,000 units

An alternative to FIFO is the weighted average cost method of inventory valuation which calculates a weighted average cost of units produced from both opening inventory and units introduced in the current period. By this method no distinction is made between units of opening inventory and new units introduced to the process during the accounting period. The cost of opening inventory is added to costs incurred during the period, and completed units of opening inventory are given a full equivalent value to each unit of production.

4.3.5 Process Costing, Joint Products and By-products

分步成本法中的聯合產品和副產品。

Joint products are two or more products separated in a process, each of which has a significant value compared to the other. Joint products have a substantial sales value. They often require further processing before they are ready for sale. Joint products arise, for example, in the oil refining industry where diesel fuel, petrol, paraffin and lubricants are all produced from the same process.

A by-product is an incidental product from a process which has an insignificant value compared to the main product. The distinguishing feature of a by-product is its relatively low sales value in comparison to the main product. In the timber industry, for example, by-products include sawdust, small off-cuts and bark.

What exactly separates a joint product from a by-product?

a) A joint product is regarded as an important saleable item, and so it should be separately costed. The profitability of each joint product should be assessed in the cost accounts.

b) A by-product is not important as a saleable item, and whatever revenue it earns is a「bonus」for the organisation. Because of their relative insignificance, by-products are not separately costed.

(1) Dealing with Common Costs

共同成本的處置。處理共同成本的方法有物理上的計量方法,如按重量、體積等分配共同成本;也可以採用相對銷售價值的分配方法,如按銷售價值分配共同成本。

The point at which joint products and by-products become separately identifiable is known as the split-off point or separation point. Costs incurred prior to this point of separation are common or joint costs, and these costs need allocating in some manner to each of the joint products. The main methods of apportioning joint costs, each of which can produce significantly different results are as follows.

a) Physical measurement.

b) Relative sales value apportionment method; sales value at split-off point.

Example 4.12

Two products (W and X) are created from a joint process. Both products can be sold immediately after split-off. There are no opening inventories or work in progress. The following information is available for last period:

Total joint production costs $776,160

Product	Production units	Sales units	Selling price per unit
W	12,000	10,000	$10
X	10,000	8,000	$12

Using the sales value method of apportioning joint production costs, what was the value of the closing inventory of product X for last period?

A. $310,464
B. $388,080
C. $155,232
D. $77,616

Solution:
The correct answer is D.
Sales value of production:
W (12,000 units × $10) $120,000
X (10,000 units × $12) $120,000

Joint production costs will be apportioned equally between the two products as the sales value of production is the same for each product.

Joint production costs allocated to X
 = $776,160 ÷ 2 = $388,080

Value of closing inventory of product X
 = 2,000 ÷ 10,000 × $388,080 = $77,616

(2) Dealing with By-products

A by-product has some commercial value and any income generated from it may be treated as follows.

a) Income (minus any post-separation further processing or selling costs) from the sale of the byproduct may be added to sales of the main product, thereby increasing sales turnover for the period.

b) The sales of the by-product may be treated as a separate, incidental source of income against which are set only post-separation costs (if any) of the by-product. The revenue would be recorded in the income statement as other income'.

c) The sales income of the by-product may be deducted from the cost of production or cost of sales of the main product.

d) The net realizable value of the by-product may be deducted from the cost of production of the main product. The net realizable value is the final saleable value of the by-product minus any post-separation costs. Any closing inventory valuation of the main product or joint products would therefore be reduced.

The choice of method (a), (b), (c) or (d) will be influenced by the circumstances of production and ease of calculation, as much as by conceptual correctness. The method you are most likely to come across in examinations is method (d). An example will help to clarify the distinction between the different methods.

Example 4.13

Which of the following statements is/are correct?

(i) A by-product is a product produced at the same time as other products which has a relatively low volume compared with the other products.

(ii) Since a by-product is a saleable item it should be separately costed in the process account, and should absorb some of the process costs.

(iii) Costs incurred prior to the point of separation are known as common or joint costs.

A. (i) and (ii)   B. (i) and (iii)
C. (ii) and (iii)   D. (iii) only

Solution:

The correct answer is D.

Statement (i) is incorrect because the value of the product described could be relatively high even though the output volume is relatively low. This product would then be classified as a joint product.

Statement (ii) is incorrect. Since a by-product is not important as a saleable item, it is not separately costed and does not absorb any process costs.

Statement (iii) is correct. These common or joint costs are allocated or apportioned to the joint products.

4.4 Service Costing

4.4.1 Introduction
服務成本法或職能成本法。

Service costing (or function costing) is a costing method concerned with establishing the costs, not of items of production, but of services rendered.

Service costing is used in the following circumstances.

(a) A company operating in a service industry will cost its services, for which sales revenue will be earned; Examples are electricians, car hire services, road, rail or air transport services and hotels.

(b) A company may wish to establish the cost of services carried out by some of its departments; For example, the costs of the vans or lorries used in distribution, the costs of the computer department, or the staff canteen.

The main problem with service costing is the difficulty in defining a realistic cost unit that represents a suitable measure of the service provided. Frequently, a composite cost unit may be deemed more appropriate. Hotels, for example, may use the 「occupied bed-night」 as an appropriate unit for cost ascertainment and control.

Typical cost units used by companies operating in a service industry are shown below.

Service	Cost unit
Road, rail and air transport services	Passenger/mile or kilometer, ton/mile, tonne/kilometer
Hotels	Occupied bed-night
Education	Full-time student
Hospitals	Patient
Catering establishment	Meal served

Example 4.14

The following information is available for a hotel company for the latest thirty day period.

Number of rooms available per night　　40
Percentage occupancy achieved　　　　65%
Room servicing cost incurred　　　　　$3,900

What was the room servicing cost per occupied room-night last period, to the nearest cent?

A. $3.25　　　　B. $5.00
C. $97.50　　　D. $150.00

Solution:
The correct answer is B.
Number of occupied room-nights
= 40 rooms×30 nights×65% = 780

Room servicing cost per occupied room-night
= $3,900÷780 = $5.00

4.4.2 Characteristics of Services

服務成本的特徵,即同時性、無形性、同質性和非持久性。

Specific characteristics of services
a) Simultaneity.
b) Intangibility.
c) Heterogeneity.
d) Perishability.
Consider the service of providing a haircut.
a) The production and consumption of a haircut are simultaneous, and therefore it cannot be inspected for quality in advance, nor can it be returned if it is not what was required.
b) A haircut is heterogeneous and so the exact service received will vary each time: not only will two hairdressers cut hair differently, but a hairdresser will not consistently deliver the same standard of haircut.
c) A haircut is intangible in itself, and the performance of the service comprises many other intangible factors, like the music in the salon, the personality of the hairdresser, the quality of the coffee.
d) Haircuts are perishable, that is, they cannot be stored. You cannot buy them in bulk, and the hairdresser cannot do them in advance and keep them stocked away in case of heavy demand.
The incidence of work in progress in service organisations is less frequent than in other types of organisation.

Example 4.15

Which of the following is not a characteristic of service costing?
A. High level of direct costs as a proportion of total costs
B. Intangibility of output
C. Use of composite cost units
D. Can be used for internal services as well as external services

Solution:
The correct answer is A.

For most services it is difficult to identify many attributable direct costs. A high level of indirect costs must be shared over several cost units, therefore option A is not a characteristic of service costing.

4.4.3 Service Cost Analysis

Service cost analysis should be performed in a manner which ensures that the following objectives are attained.

a) Planned costs should be compared with actual costs. Differences should be investigated and corrective action taken as necessary.

b) A cost per unit of service should be calculated.

If each service has a number of variations (such as maintenance services provided by plumbers, electricians and carpenters) then the calculation of a cost per unit of each service may be necessary.

c) The cost per unit of service should be used as part of the control function.

For example, costs per unit of service can be compared, month by month, period by period, year by year and so on and any unusual trends can be investigated.

d) Prices should be calculated for services being sold to third parties.

The procedure is similar to job costing. A mark-up is added to the cost per unit of service to arrive at a selling price.

e) Costs should be analyzed into fixed, variable and semi-variable costs to help assist management with planning, control and decision making.

(1) Service Cost Analysis in Internal Service Situations

Service department costing is also used to establish a specific cost for an internal service which is a service provided by one department for another, rather than sold externally to customers eg. canteen, maintenance. Transport costs and canteen costs are service costs. 「Transport costs」 is a term used here to refer to the costs of the transport services used by a company, rather than the costs of a transport organization, such as a rail network. Another example of service costing is the cost of a company's canteen services. A feature of canteen costing is that some revenue is earned when employees pay for their meals, but the prices paid will be insufficient to cover the costs of the canteen service.

The company wills subsidies the canteen and a major purpose of canteen costing is to establish the size of the subsidy.

Example 4.16

Last year, Bryan Air carried excess baggage of 250,000kg over a distance of 7,500 km at a cost of $3,750,000 for the extra fuel.
What is the cost per kg-km?
A. $0.002 per kg-km B. $2.00 per kg-km
C. $33.33 per kg-km D. $500.00 per kg-km

Solution:
The correct answer is A.
First we calculate the total number of kg-km.
Kg × km taken = 250,000kg × 7,500 km
             = 1,875,000,000 kg-km
Cost per kg-km = $3,750,000 ÷ 1,875,000,000
             = $0.002 per kg-km

(2) Service Cost Analysis in Service Industry Situations

Service companies, such as eBay, Verizon, Bank, Education and transport organization. Service companies need to know which services are most profitable, and that means evaluating both revenues and costs. Knowing the cost per service helps managers to set the price of each and then to calculate operating income.

Example 4.17

Which of the following would be considered a service industry?
(i) An airline company
(ii) A railway company
(iii) A firm of accountants
A. (i) and (ii) only
B. (i) and (iii) only
C. (i), (ii) and (iii)
D. (ii) and (iii) only

Solution:
The correct answer is C.
An airline company, a railway company and a firm of accountants are all considered to be service industries.

QUESTIONS:

1. A company operates a job costing system. Job 812 requires $60 of direct materials, $40 of direct labour and $20 of direct expenses. Direct labour is paid $8 per hour. Production overheads are absorbed at a rate of $16 per direct labour hour and non-production overheads are absorbed at a rate of 60% of prime cost.
What is the total cost of Job 812?
A. $240
B. $260
C. $272
D. $320

2. Which one of the following statements is incorrect?
A. Job costs are collected separately, whereas process costs are averages
B. In job costing the progress of a job can be ascertained from the materials requisition notes and job tickets or time sheet
C. In process costing information is needed about work passing through a process and work remaining in each process
D. In process costing, but not job costing, the cost of normal loss will be incorporated into normal product costs

3. A company uses process costing to value output. During the last month the following information was recorded:
Output: 2,800 kg valued at $7.50/kg
Normal loss: 300 kg which has a scrap value of $3/kg
Actual gain: 100 kg
What was the value of the input?
A. $22,650
B. $21,900
C. $21,600
D. $21,150

4. In process costing, if an abnormal loss arises the process account is generally:
A. Debited with the scrap value of the abnormal loss units
B. Debited with the full production cost of the abnormal loss units
C. Credited with the scrap value of the abnormal loss units
D. Credited with the full production cost of the abnormal loss units

5. ABC manufactures product X in a single process. Normal loss (scrap) in the process is 10% of output and scrapped units can be sold off for $4/unit.
In period 8 there was no opening inventory and no closing inventory. Process costs of direct materials, direct labour and production overheads totaled $184,800. Input to the process in the

month was 13,200 units.
What was the cost/unit produced?
A. $12.50
B. $15.00
C. $15.15
D. $15.40

6. A company uses process costing to value its output. The following was recorded for the period:

Input materials	2,000 units at $4.50 per unit
Conversion costs	$13,340
Normal loss	5% of input valued at $3 per unit
Abnormal loss	150 units

There were no opening or closing inventories.
What was the valuation of one unit of output?
A. $11.80
B. $11.60
C. $11.20
D. $11.00

7. A company that operates a process costing system had work-in-progress at the start of last month of 300 units (valued at $1,710) that were 60% complete in respect of all costs. Last month a total of 2,000 units were completed and transferred to the finished goods warehouse. The cost per equivalent unit for costs arising last month was $10. The company uses the FIFO method of cost allocation.
What was the total value of the 2,000 units transferred to the finished goods warehouse last month?
A. $19,910
B. $20,000
C. $20,510
D. $21,710

8. A company operates a job costing system.
Job number 605 requires $300 of direct materials and $400 of direct labour. Direct labour is paid at the rate of $8 per hour. Production overheads are absorbed at a rate of $26 per direct labour hour and non-production overheads are absorbed at a rate of 120% of prime cost.
What is the total cost of job number 605?
A. $2,000
B. $2,400
C. $2,840
D. $4,400

9. **A hotel calculates a number of statistics including average cost per occupied bed per day.**
The following information is provided for a 30-day period.

	Rooms with twin beds	Single rooms
Number of rooms in hotel	260	70
Number of rooms available to let	240	40
Average number of rooms occupied daily	200	30
Number of guests in period	6,450	
Average length of stay	2 days	
Payroll costs for period	$100,000	
Cost of cleaning supplies in period	$5,000	
Total cost of laundering in period	$22,500	

The average cost per occupied bed per day for the period is:
A. $9.90
B. $9.88
C. $7.20
D. $8.17

10. **A hotel calculates a number of statistics including average room occupancy.**
Average room occupancy is calculated as the total number of rooms occupied as a percentage of rooms available to let.
The following information is provided for a 30-day period.

	Rooms with twin beds	Single rooms
Number of rooms in hotel	260	70
Number of rooms available to let	240	40
Average number of rooms occupied daily	200	30

The average room occupancy is:
A. 69.7%　　　B. 82.1%　　　C. 82.7%　　　D. 84.8%

Chapter 5　Activity Based Costing and Other Cost Management Tools

Learning Objectives

After the study of this chapter, you should be able to:
a) Explain activity based costing (ABC), target costing, life cycle costing and total quality management as alternative cost management techniques.
b) Differentiate ABC, Target costing and life cycle costing from the traditional costing techniques.

5.1　Activity Based Costing (ABC)

5.1.1　Defining Activity Based Costing (ABC)

作業成本法（ABC）是傳統的成本計算技術的發展，它認為成本是生產產品的作業產生的，因此間接費用應按作業進行分配，進而按產品消耗的作業多少分配到產品上去。

Activity based costing (ABC) is an alternative to the traditional costing techniques. ABC involves the identification of the factors (cost drivers) which cause the costs of an organization's major activities. Support overheads are charged to products on the basis of their usage of the factor causing the overheads (an activity).

With the advent of advanced manufacturing technology, overheads are likely to be far more important. Traditional costing techniques, which assume that all products consume all resources in proportion to their production volumes, tend to allocate too great a proportion of overheads to high volume products (which cause relatively little diversity and hence use fewer support services) and too small a proportion of overheads to low volume products (which cause greater diversity and therefore use more support services). So it is difficult to justify the use of production volume as the basis for allocating overheads. It is against this background that activity based costing (ABC) has emerged.

The major ideas behind activity based costing are as follows.
a) Activities cause costs. Activities include ordering, materials handling, machining, assembly, production scheduling and dispatching.

b) Producing products creates demand for the activities.

c) Costs are assigned to a product on the basis of the product's consumption of the activities.

5.1.2 Calculating Product Costs Using ABC

An ABC system operates as follows.

Step1: Identify an organization's major activities.

Step2: Identify the factors which determine the size of the costs of an activity/cause the costs of an activity. these are known as cost drivers.

Step3: Collect the cost of each activity into what are known as cost pools.

Step4: Charge support overheads to products on the basis of their usage of the activity. A product's usage of an activity is measured by the number of the activity's cost driver it generates.

Example 5.1

ABC Company makes and sells a number of products. Products M and N are products for which market prices are available at which S can obtain a share of the market as detailed below. Estimated data for the forthcoming period is as follows.

a) Product data is shown in table 5.1 as following:

Table 5.1   **Product Data**

	Product M	Product N	Product P
Production/sales (units)	5,000	10,000	40,000
	$'000	$'000	$'000
Total direct material cost	80	300	2,020
Total direct labour cost	40	100	660

b) Variable overhead cost is $1,500,000 of which 40% is related to the acquisition, storage and use of direct materials and 60% is related to the control and use of direct labour.

c) It is current practice in ABC company to absorb variable overhead cost into product units using overall company wide percentages on direct material cost and direct labour cost as the absorption bases.

Required:

Prepare estimated unit product costs for product M and N where variable overhead is charged to product units using an activity based costing approach where cost drivers have been estimated for material and labour related overhead costs as follows, please look at table 5.2.

Table 5.2   **Material and Labour Related Overhead Costs**

	Product M	Product N	Product P
Direct material related overheads—cost driver is material bulk. The bulk proportions per unit are	4	1	1.5
Direct labour related overheads—cost driver is number of labour operations. Labour proportions per unit are	6	1	2

Solution:

Overhead per cost driver of material-related overhead:

($1,500,000 × 40%) ÷ (4 × 5,000 + 1 × 10,000 + 1.5 × 40,000)

= $600,000 ÷ 90,000

= $6.67

Overhead per cost driver of labour-related overhead:

($1,500,000 × 60%) ÷ (6 × 5,000 + 1 × 10,000 + 2 × 40,000)

= $900,000 ÷ 120,000

= $7.50

Product costs per unit for product M and product N using ABC approach is shown in table 5.3 as following:

Table 5.3   **Product Costs Per Unit for Product M and Product N Using ABC Approach**

	Product M	Product N
Direct material cost per unit ($80,000/5,000, $300,000/10,000)	16.00	30.00
Direct labour cost per unit ($40,000/5,000, $100,000/10,000)	8.00	10.00
Material-related OH per unit ($6.67×4, $6.67×1)	26.68	6.67
Labour-related OH per unit ($7.50×6, $7.50×1)	45.00	7.50
Production cost per unit	95.68	54.17

5.2 Target Costing

5.2.1 Defining Target Costing

目標成本通過市場價格減去期望的利潤來確定。

Target Costing involves setting a target cost by subtracting a desired profit margin from a competitive market price.

In today's competitive market, companies must continually redesign their products with the result that product life cycles have become much shorter. The planning, development and design stage of a product is therefore critical to a company's cost management process. Considering possible cost reductions at this stage of a product's life cycle is now one of the most important issues facing management accountants in industry. Here are some examples of decisions made at the design stage which directly impact on the cost of a product.

a) The number of different components.
b) Whether the components are standard or not.
c) The ease of changing over tools.

Target costing have been developed as a response to the problem of controlling and reducing costs over the product life cycle.

Target costing has its greatest impact at the design stage because a large percentage of a product's life cycle costs are determined by decisions made early in its life cycle. So the technique requires managers to change the way they think about the relationship between cost, price and profit.

a) Traditionally the approach is to develop a product, determine the production cost of that product, set a selling price, with a resulting profit or loss. This cost plus pricing is a「bottom up」approach where you start with full production costs and add a margin to arrive at a selling price. However it does not take account of market conditions.

b) The target costing approach is to develop a product, determine the market selling price and desired profit margin, with a resulting cost which must be achieved. Target costing is a「top down」method and starts with a price that take account of market conditions and deducts a desired margin to arrive at a target cost.

5.2.2 Implementing Target Costing

In a general way, the implementation of the target costing process is the following steps.

a) Determine a product specification of which an adequate sales volume is estimated.
b) Set a selling price at which the company will be able to achieve a desired market.
c) Estimate the required profit based on return on sales or return on investment.
d) Calculate the target cost = Target selling price − Target profit.
e) Compile an estimated cost for the product based on the anticipated design specification and current cost levels.
f) Calculate cost gap = Estimated cost − Target cost.

g) Make efforts to close the gap. This is more likely to be successful if efforts are made to 「design out」 costs prior to production, rather than to 「control out」 costs during the production phase.

h) Negotiate with the customer before making the decision about whether to go ahead with the project.

When a product is first manufactured, its target cost may well be much lower than its currently-attainable cost, which is determined by current technology and processes. Management can then set benchmarks for improvement towards the target costs, by improving technologies and processes.

Even if the product can be produced within the target cost, target costing can also be applied throughout the entire life cycle. Once the product goes into production target costs will therefore gradually be reduced. These reductions will be incorporated into the budgeting process. This means that cost savings must be actively sought and made continuously over the life of the product.

5.3 Life Cycle Costing

5.3.1 Defining Life Cycle Costing

壽命週期成本將產品的成本範圍擴展到產品的整個壽命週期,不僅考慮產品的生產成本,而且考慮產品的研發成本以及售後的服務成本、維修成本。

Life cycle costing tracks, accumulates costs and revenues attributable to each product over the entire product life cycle. A product life cycle can be divided into four phases.

a) Introduction. The product is introduced to the market. Potential customers will be unaware of the product or service, and the company may have to spend further on advertising to bring the product or service to the attention of the market.

b) Growth. The product gains a bigger market as demand builds up. Sales revenues increase and the product begins to make a profit.

c) Maturity. Eventually, the growth in demand for the product will slow down and it will enter a period of relative maturity. It will continue to be profitable. The product may be modified or improved, as a means of sustaining its demand.

d) Decline. At some stage, the market will have bought enough of the product and it will therefore reach 「saturation point」. Demand will start to fall. Eventually it will become a loss-maker and this is the time when the company should decide to stop selling the product or service.

A product's life cycle costs are incurred from its design stage through development to market launch, production and sales, and finally to its eventual withdrawal from the market. The component elements of a product's cost over its life cycle could therefore include the following.

a) Research & development costs.
b) The cost of purchasing any technical data required.
c) Retirement and disposal cost.
d) Training costs (including initial operator training and skills updating).
e) Production costs.
f) Distribution costs and marketing costs.
g) Inventory cost (holding spare parts, warehousing and so on).

Life cycle costs can apply to services, customers and projects as well as to physical product.

5.3.2 The Value of Life Cycle Costing

Traditional cost accumulation systems are based on the financial accounting year and tend to dissect a product's life cycle into a series of 12-month periods. This means that traditional management accounting systems do not accumulate cost over a product's entire life cycle and do not therefore assess a product's profitability over its entire life. Instead they do it on a periodic basis.

Life cycle costing, on the other hand, tracks and accumulates actual costs and revenues attributable to each product over the entire product life cycle. Hence the total profitability of any given product can be determined.

Traditional cost accumulation systems usually total all non-production costs and record them as a period expense. With life cycle costing, non-production costs are traced to individual products over complete life cycles. So life cycle costing has the value as the following.

a) The total of these costs for each individual product can therefore be reported and compared with revenues generated in the future.
b) The visibility of such costs is increased.
c) Individual product profitability can be more fully understood by attributing all costs to products.
d) As a consequence, more accurate feedback information is available on the company's success or failure in developing new products. In the competitive environment that is vital.

5.4 Total Quality Management (TQM)

全面質量管理(TQM)。

Total quality management is the process of applying a zero defect philosophy to the management of all resources and relationships within an organization as a means of developing and sustaining a culture of continuous improvement which focuses on meeting customer expectations.

The basic principles of TQM include the following:

a) The cost of preventing mistakes is less than the cost of correcting them once they occur. The aim should therefore be to get things right first time.

b) Dissatisfaction with the status quo: the belief that it is always possible to improve and so the aim should be to 「get it more right next time」.

The process of the management of quality:

a) Establishing standards of quality for a product or service.

b) Establishing procedures or production methods which ought to ensure that these required standards of quality are met in a suitably high proportion of cases.

c) Monitoring actual quality.

d) Taking control action when actual quality falls below standard.

QUESTIONS:

1. Which one of the following is an advantage of activity based costing?
A. It provides more accurate product costs
B. It is simple to apply
C. It is a form of marginal costing and so is relevant to decision making
D. It is particularly useful when fixed overheads are very low

2. Quality control costs can be categorized into internal and external failure costs, inspection costs and prevention costs. In which of these four classifications would the following costs be included?
(i) The costs of a customer service team
(ii) The cost of equipment maintenance
(iii) The cost of operating test equipment

	Customer service team	Equipment maintenance	Test equipment
A.	Prevention costs	Inspection costs	Internal failure costs

B. Prevention costs	Internal failure costs	Inspection costs
C. External failure costs	Internal failure costs	Prevention costs
D. External failure costs	Prevention costs	Inspection costs

3. In the context of quality costs, customer compensation costs and test equipment running costs would be classified as:

	Customer compensation costs	Test equipment running costs
A.	Internal failure costs	Prevention costs
B.	Internal failure costs	Appraisal costs
C.	External failure costs	Appraisal costs
D.	External failure costs	Prevention costs

4. The selling price of product K is set at $450 for each unit.
If the company requires a return of 20% in the coming year on product K, the target cost for each unit for the coming year is:
A. $300
B. $360
C. $400
D. $450

5. In calculating the life cycle costs of a product, which of the following items would be excluded?
(ⅰ) Planning and concept design costs
(ⅱ) Preliminary and detailed design costs
(ⅲ) Testing costs
(ⅳ) Production costs
(ⅴ) Distribution and customer service costs
A. (ⅲ)
B. (ⅳ)
C. (ⅴ)
D. None of them

6. As part of a process to achieve a target cost, GYE Inc is interviewing prospective customers to determine why they would buy the product and how they would use it.
What term best describes this process?
A. Value analysis
B. Operational research
C. TQM
D. Lifecycle costing

7. A customer returns a faulty product to a firm for repair under a warranty scheme. The firm operates a total quality management system.
Which of the following best describes the cost of the repair?
A. An internal failure cost
B. An external failure cost
C. An appraisal cost
D. A prevention cost

Chapter 6　Cost Behavior and Cost-Volume-Profit (CVP) Analysis

Learning Objectives

After the study of this chapter, you should be able to:
a) Explain and illustrate the nature of variable, fixed and mixed (semi-variable, stepped-fixed) costs. Use the high-low method to separate semi-variable costs. Use variable, fixed and semi-variable costs in cost analysis.
b) Explain Cost-Volume-Profit (CVP) Analysis. Calculate contribution per unit and the contribution/sales ratio.
c) Explain the concept of break-even and margin of safety, use contribution per unit and contribution/sales ratio to calculate break-even point and margin of safety.
d) Analyse the effect on break-even point and margin of safety of changes in selling price and costs.
e) Use contribution per unit and contribution/sales ratio to calculate the sales required to achieve a target profit.
f) Construct break-even and profit/volume charts for a single product or business.
g) Calculate breakeven sales or target sales in units for each product of sales mix.

6.1　Cost Behavior

0.1.1　Introduction to Cost Behavior

　　成本習性是一個成本項目的總額隨業務量水平的變化而變化的方式,它最終可以將所有的成本分為固定成本和變動成本。企業管理者據此可通過分析不同業務量水平的成本、收入進行經營決策、經營控制和預算管理。

Managers need to know how a business's costs are affected by changes in its level of activity. Cost behavior is the way in which costs are affected by changes in the level of activity. Costs by cost behavior are classified four types of costs:
a) Variable cost.
b) Fixed cost.
c) Mixed cost/Semi-variable cost/Semi-fixed cost.
d) Step cost.

Activity level normally refers to the amount of work done or the volume of production. The level of activity may refer to one of the following:
a) Labour hours produced.
b) Machine hours produced.

c) Number of units produced.
d) Number of items sold.
e) Value of items sold.
f) Number of invoices issued.
g) Number of units of electricity consumed.

The basic principle of cost behavior is that as the level of activity rises, costs will usually rises. The cost to product 500 units of output will be more the cost to product 400 units.

Management decisions will often be based on how costs and revenues vary at different activity levels. So knowledge of cost behavior is obviously essential for decision-making, control, and budgeting.

6.1.2 Cost Behavior Application

所有的成本項目可分為變動成本、固定成本和混合成本,通過將混合成本分解為固定成本和變動成本,則可建立總成本的成本習性模型並進行成本分析、預測。

總成本習性模型如下:
總成本(TC)＝固定成本＋單位變動成本×業務量水平
TC = FC+ VC/u× Units

混合成本的分解主要有高低點法和最小二乘迴歸法。

(1) Cost Estimation

It is generally assumed that costs are one of the followings.
a) Variable cost.
b) Fixed cost.
c) Mixed cost.

Cost accountants tend to separate mixed cost into variable and fixed elements. They therefore generally tend to treat costs as either fixed cost or variable cost and to estimate total costs by fixed cost plus variable cost in some activity level. The formula is as follows.

$$\text{Total costs} = FC + VC/u \times \text{Activity level}$$

There are several methods for identifying the fixed and variable element of mixed costs. Each method is only an estimate, and each will product different results. The principal methods of separating mixed cost into variable and fixed elements are the high-low method and the least squares regression method.

(2) High-low Method

The High-low method is an easy method to separate mixed cost into variable and fixed elements. This method requires you to identify the highest and lowest activity levels (i.e. labor hours, machine hours, production unit etc.) over a period of time and their corresponding total cost figures. Then using this information, you should complete the following steps.

a) Calculate the variable cost per unit.

$$VC/u = \frac{\text{Total cost at highest activity level} - \text{Total cost at lowest activity level}}{\text{Total unis at highest activity level} - \text{Total units at lowest activity level}}$$

b) Calculate the total fixed costs.

Total FC = Total cost at highest activity level− VC/u×Total units at highest activity level

or

Total FC = Total cost at lowest activity level− VC/u×Total units at lowest activity level

c) Create and use an equation to show the behavior of a mixed cost.

Total mixed cost = Total FC + VC/u × units

Example 6.1　　FK Company has recorded the following total costs during the six years, please look at table 6.1 as following:

Table 6.1　　FK Company's Total Costs During the Five Years

Year	2010	2011	2012	2013	2014	2015
Output volume Units	65,000	60,000	80,000	90,000	85,000	75,000
Total cost $	145,000	140,000	162,000	170,000	165,000	158,000

Required:

Calculate the total cost that should be expected in 2016 if output is 82,000 units.

Solution:

a) $VC/u = \dfrac{170,000-140,000}{90,000-60,000} = \$1/u$

b) Total FC = 170,000 − ($1/u) × 90,000
　　　　　 = $80,000

c) Therefore the total costs in 2016 for output of 82,000 units are as follows.

　　　$80,000+ ($1/u) ×82,000 = $162,000

Example 6.2　　A company has the following total costs at three activity levels.

Activity level (units)　 8,000　　12,000　　15,000
Total cost　　　　　$204,000　$250,000　$274,000

Variable cost per unit is constant within this activity range and there is a step up of 10% in the total fixed costs when the activity level exceeds 11,000 units.

What is the total cost at an activity level of 10,000 units?

Solution:

a) $VC/u = \dfrac{274,000-250,000}{15,000-12,000} = \$8/u$

b) Total fixed cost above 11,000 units
　　= 274,000 −(15,000× $8/u)
　　= $154,000

Total fixed cost below 11,000 units
= 154,000 ÷ (1+10%)
= $140,000
c) Therefore the total cost for 10,000 units
= 140,000 + (10,000 × $8/u)
= $220,000

Example 6.3

Use the high-low method to calculate (a) variable cost per unit, (b) total fixed costs, and to determine (c) the total cost equation.

month	activity	Total cost($)
September	50,000	85,000
October	65,000	100,000
November	49,000	80,000
December	79,000	101,000

Solution:
Select the activity with the highest and lowest amounts. Choose November and December.
a) The variable cost per unit
= [$101,000 − $80,000] ÷ [79,000 − 49,000]
= $0.70 per unit
b) Plug the variable cost from step (a) and the activity and total cost from either the high or the low point into the total cost equation, to solve for fixed costs (FC):
TC = 101,000 = 0.7×79,000 + FC
FC = $45,700
c) The total cost equation is then:
TC = 45,700 + 0.70×79,000

(3) Least Squares Regression Method

The least squares regression method uses all the available cost and activity data to form a regression line that minimizes the deviations to separate mixed costs into variable and fixed components.

The formulas used to calculate least squares are:

$$b = \frac{n(\sum xy) - (\sum x)(\sum y)}{n(\sum x^2) - (\sum x)^2}$$

$$a = \frac{(\sum y) - b(\sum x)}{n}$$

Where
n = Number of observations

x = The level of activity
y = The total mixed cost
a = The total fixed cost
b = The variable cost per unit of activity

Example 6.4

The production and total cost information relates to a single product organization for the last six months, please look at table 6.2 as following:

Table 6.2　　　　**Production and Total Cost Information**

Month	1	2	3	4	5	6
Production units	1,200	900	1,500	2,500	2,900	2,200
Total cost($)	65,000	58,200	68,500	87,500	95,600	84,700

The variable cost per unit and fixed cost is constant up to a production level of 3,000 units per month.

What is the total cost for next month when 2,800 units are produced?

Solution (look at table 6.3 as follows):

Table 6.3

Month—n	1	2	3	4	5	6	Σ
Production units—x	1,200	900	1,500	2,500	2,900	2,200	11,200
Total cost($)—$y$	65,000	58,200	68,500	87,500	95,600	84,700	459,500
xy	78,000,000	52,380,000	102,750,000	218,750,000	277,240,000	186,340,000	915,460,000
x^2	1,440,000	810,000	2,250,000	6,250,000	8,410,000	4,840,000	24,000,000

b = (6×915,460,000−11,200×459,500)÷(6×24,000,000−11,200^2)
　= 18.66
a = (459,500−18.66×11,200)÷6
　= 41,751.33
The total cost for next month when 2,800 units are produced:
　　= 41,751.33+18.66×2,800 = 93,999.33

6.2　Basic CVP Analysis

　　本量利分析(CVP Analysis)是基於成本習性通過收入、成本與業務量的數量關係來分析經營活動中成　　Cost-Volume-Profit Analysis is an abbreviation of the Analysis of the relationship between the Cost, the production (or sales), and the profit, also calls the CVP Analysis. The formation and development of this analysis method is on the basis of the classification of the cost according cost

本、銷售收入、利潤三者之間相互依存關係的。

6.2.1 The Basic Assumptions of CVP Analysis

相關範圍假設包括時期範圍假設和業務範圍假設。時期範圍假設是指固定成本的固定不變和變動成本的成比例變動是對於一定的時期範圍而言的,當時間超過這個時期範圍,它們的這種性質就會發生變化,固定成本的內容及其總額、單位變動成本的內容及其金額都將變動。同樣,業務範圍假設是指固定成本的固定不變和變動成本的成比例變動是對於一定的業務量範圍而言的,當業務超過這個業務量範圍,它們的這種性質就會發生變化,固定成本的內容及其總額、單位變動成本的內容及其金額都將變動。

產銷平衡假設。

behavior, which main research is the relationship between sales, prices, cost and profit.

The CVP Analysis is based on the following assumptions.
(1) **Relevant Range Hypothesis**
Relevant range hypothesis includes 「period hypothesis」 and 「business hypothesis」. 「Period hypothesis」 refers to the assumption that the fixity of the fixed costs and the proportion to the change of the variable cost only work in a certain period. Beyond the period the nature of them will change, and so will the content of the fixed cost and its value, as well as the content of the variable unit cost and its total amount. Likewise, 「business hypothesis」 refers to the assumption that the fixity of the fixed costs and the proportion to the change of the variable cost work only within a certain range of business volume, when the volume exceeds the range the nature of them will change, as well as the content of the fixed cost and its value, and the content of the unit variable cost and total amount.

(2) **Model of Linear Hypothesis**
a) Fixed cost constant hypothesis;
　　Volume change—FC constant
b) Variable cost and volume is completely linearity hypothesis;
　　VC = VC/u×Volume
c) Total sale revenues and sales quantity has been completely linearity hypothesis.
　　Total Revenues = Sale Price × Sale Quantity
Thus, the selling prices and variable costs are constant per unit at all volumes of sales, and the fixed costs remain fixed at all levels of activity.

(3) **Production and Marketing Balance Assumption**
The core of CVP analysis is to analyze the relation between income and cost. The changes in production volume are likely to affect fixed costs, or the variable costs, and this influence will certainly affect the contrast relationship between income and cost. Therefore in CVP analysis which stands in the angle of the sales quantity, the balanced relationship between production volume and sale volume must be assumed.

品種結構不變假設。本假設是指在一個多產品品種生產和銷售的企業中，各種產品的銷售收入占總收入的比重不會發生變化。由於多品種條件下各種產品的獲利能力一般會有所不同，有些差異還比較大。例如，企業產銷的品種結構發生較大的變動，勢必導致預計利潤與實際利潤之間出現較大的「非預計性」差異。

(4) Multi-product Structure Constant Hypothesis

In this hypothesis it is assumed that in a business of various productions and sales the proportion of every product in the total sales revenue will remain fixed. This is because that the profitability of every product usually varies under the condition of multi-production, thus a larger change in the structure of enterprise production would lead to an 「unexpected」 difference between the expected profit and the actual profit.

6.2.2 The Basic Equation of CVP Analysis

CVP analysis is based on marginal costing principles. Its greatest value as a technique is for estimating and providing information for decision making. The value of CVP analysis depends on making reliable estimates of variable costs and fixed costs. Where some costs are semi-fixed and semi-variable, these should be divided into a variable cost element and a fixed cost element. Based on the above assumptions we can establish the function relation between cost, quantity and profit.

Operating profit
= Total sale revenues − Total costs
= Total sales revenue − Variable costs − Fixed costs
= Sales quantity × (Sale price − Variable costs per unit) − Fixed costs
or, Operating profit
= TR − TC = TR − VC − FC
= SQ × (SP − VC/u) − FC

邊際貢獻（Contribution）是本量利分析中的一個關鍵概念，等於銷售收入減去變動成本的差額，彌補固定成本之後即實現利潤。

In CVP analysis, contribution is a key concept, because if we assume a constant variable cost per unit and the same selling price at all volumes of output, the contribution per unit is a constant value.

Unit contribution
= Selling price per unit − Variable costs per unit
Total contribution
= Volume of sales in units × (Unit contribution)
Or, = Total sales revenue − Total variable cost
Or, = Total sales revenue × Contribution/Sales ratio

Contribution/Sales ratio or C/S ratio
= Contribution per unit/Sales price per unit
Or, = Total contribution/Total sales revenue

Applications of CVP analysis include:
a) estimating future profits.
b) calculating the break-even point for sales.
c) analyzing the margin of safety in the budget.
d) calculating the volume of sales required to achieve a target profit.
e) deciding on a selling price for a product.

Example 6.5

A company has a single product. The following budgeted information relates to a period:

Sales units	800,000
Sales revenue	$1,000,000
Total variable costs	$590,000
Total fixed costs	$350,000

Calculate the Operating profit, Total contribution, Unit contribution and C/S ratio that should be expected in the period.

Solution:
Operating profit
    = 1,000,000 − 590,000 − 350,000 = $60,000
Total contribution
    = 1,000,000 − 590,000 = $410,000
Unit contribution
    = 410,000 ÷ 800,000 = $0.512,5 per unit
C/S ratio
    = 410,000 ÷ 1,000,000 = 41%

6.3 Using CVP to Break-even Analysis

Break-even analysis is based on factors such as cost, revenue and profit, and the function relationship between those factors, forecasting enterprise in the case of how to break even, or the profit being equal to zero.

Break-even point refers to the business enterprise scale (sales) just to enable enterprises to achieve the break even, or the profit being equal to zero.

To be specific, the break-even analysis mainly includes three aspects of content:

a) Basic calculation model of break-even point.
b) The critical figure of target profit.
c) The influence of related factors change on the break-even point.

6.3.1 Basic Calculation Model of Break-even Point

Break-even is the volume of sales at which the business just 「breaks even」, so that it makes neither a loss nor a profit. At break-even point, total contribution is just large enough to cover fixed costs. In other words, at break-even point:

$$\text{Total contribution} = \text{Fixed costs}$$

Formulas:
Break-even point in unit:
BE (u) = Fixed costs / Contribution per unit
Break-even point in sales:
BE ($) = Fixed costs / C/S ratio

Example 6.6

A company manufactures and sells a single product which has the following cost and selling price structure:

	$/unit	$/unit
Selling price		120
Direct material	22	
Direct labour	36	
Variable overhead	14	
Fixed overhead	12	(84)
Profit		36

The fixed overhead absorption rate is based on the normal capacity of 2,000 units per month. Assume that the same amount is spent each month on fixed overheads. Budgeted sales for next month are 2,200 units.

Required:
a) What is the break-even point in units?
b) What is the break-even point in sales revenue?
c) What would be the break-even point if fixed costs went up to $30,000?
d) What would be the break-even point if fixed costs were $24,000 but unit variable costs went up to $80?

Solution:
a) Contribution per unit = 120 − (22+36+14)
 = 120−72 = 48
  so, BE(u) = (2,000×12) ÷ 48 = 500 units
b) BE($) = 500×120 = $60,000
c) If fixed costs went up to $30,000,

BE(u) = 30,000÷48 = 625 units
BE($) = 625×120 = $75,000
d) If fixed costs were $24,000 but unit variable costs went up to $80,
BE(u) = 24,000÷(120−80) = 600 units
BE($) = 600×120 = $72,000

安全邊際(MOS)是預算銷售量(額)或正常銷售量(額)和盈虧平衡點銷售量(額)的差額,它相對於預算銷售量(額)或正常銷售量(額)的百分比即安全邊際率。兩者反應企業經營的安全性,兩者的數值越大,經營安全性越高。

Margin of safety (MOS) is the difference between the budgeted sales volume and the break-even sales volume. It is simply a measurement of how far sales can fall short of budget before the business makes a loss. A large margin of safety indicates a low risk of making a loss, whereas a small margin of safety might indicate a fairly high risk of making a loss. It therefore indicates the vulnerability of a business to a fall in demand.

Formulas:
MOS (u) = Budgeted sales (u) − BE (u),
    or = Profit/Contribution per unit
MOS ($) = Budgeted sales ($) − BE ($),
    or = Profit/C/S ratio
MOS (%) = (Budgeted sales − Break even sales level)/ Budgeted sales

Example 6.7

XYZ company makes and sells a single product. The details are as follow:
Selling price $24 per unit
Variable cost $14 per unit
Budgeted sales for the year are 140,000 units.
Budgeted fixed costs for the year are $840,000.
Required:
a) Calculate the break-even point in units and $.
b) Calculate the margin of safety in units and $.
c) Calculate the margin of safety, as a percentage of budgeted sales.

Solution:
a) Contribution per unit = 24−14 = 10
So, BE (u) = 840,000÷10 = 84,000
BE ($) = 84,000×24 = 2,016,000
b) MOS (u) = 140,000 − 84,000 = 56,000
MOS ($) = 140,000×24 − 2,016,000 = 1,344,000
c) MOS (%) = (140,000 − 84,000)÷140,000 = 40%

$$\text{Or, } = (140,000 \times 24 - 2,016,000) \div (140,000 \times 24)$$
$$= 40\%$$

6.3.2 The Critical Figure of Profit and Loss

Critical figure of profit and loss is the break-even point analysis reflected in rectangular coordinate system. And critical graph according to the characteristics of the data and purpose is different, can have a variety of forms. A graphical presentation of CVP analysis can be made in either a conventional break-even chart (figure 6.1) or a profit/volume chart.

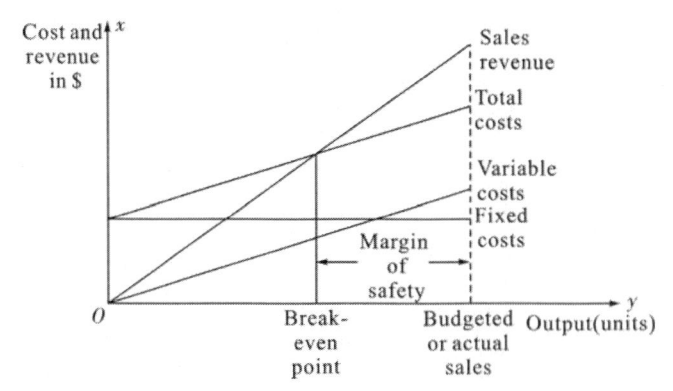

Figure 6.1 Conventional Break-even Chart

The x axis represents sales volume, in units.
The y axis represents cost and revenue, in $.

Based on the conventional break-even chart intuitive and vividly describes the relationship, specifically manifested in the following aspects:

a) In the fixed cost, variable cost per unit, under the condition of sales unit price remains the same, that is break-even point is established, the greater the sales, to achieve more profit achieved (when sales more than break-even point), or the less loss (when insufficient sales break-even point); Conversely the less profit or loss. This is the basic relationship of the critical figure of profit and loss.

b) In the case of the total cost of the established, the location of the break-even point along with the change of sales unit price reverse changes: the higher sale price, the higher sales revenue (shown in the drawing is the greater the slope of line), the lower the break-even point is; On the other hand, the higher the break-even point is.

c) The sale price unit variable costs established cases, the location of the break-even point moves according to the change of the total amount of fixed costs: the greater fixed costs (in the coordinate chart is the total cost line and the higher the intersection of the vertical axis), the higher the break-even point is; On the contrary, the lower the break-even point is.

d) The sales unit price and total fixed cost under the condition of given, the location of the break-even point along with the change of the unit variable cost moves: the higher unit variable cost (shown in the drawing is the greater the slope of the total cost line), the higher the break-even point is; On the contrary, the lower the break-even point is.

A profit/volume graph is an adaptation of a break-even chart which makes it easier to determine the profit or loss at each volume of sales, please look at figure 6.2 as following:

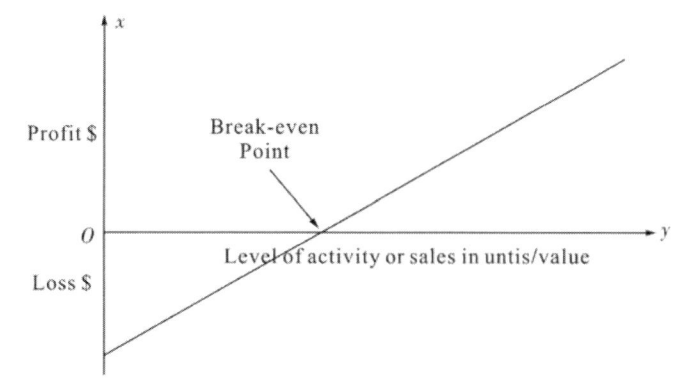

Figure 6.2 Profit Volume Graph

The x axis represents sales volume, in units or $.
The y axis represents loss or profit.
The x axis cuts the y axis at break-even point (profit=0).
Losses are plotted below the line and profits above the line.
To draw the chart, only two points need to be plotted on the graph. These can be:
a) Profit at planned or budgeted sales volume.
b) Loss at zero sales volume, which is equal to total fixed costs.

6.3.3 The Influence of Related Factors Change on the Break-even Point

Requirements in factors such as fixed cost, variable cost per unit, the sale price and sale variety structure changes can calculate new break-even point, and understand the above various factors changes direction and break-even point changes direction relationship based on the following model:

Break-even point in sale units = Fixed cost/(Unit price - Unit variable cost)

Break-even point in sale revenue = Fixed cost/Marginal contribution rate

a) Fixed costs change on the influence of the break-even point: fixed cost increase will lead to the rise of break-even point, reduce fixed costs can lead to lower break-even point.

b) Unit variable costs change on the influence of the break-even point: unit variable cost increase will lead to the rise of break-even point, the unit variable cost reduction will lead to lower break-even point.

c) Sales unit price change on the influence of the break-even point: the rise in the price of sales leads to lower break-even point, sales falling prices will lead to the rise of break-even point.

d) Variety structure change on the influence of the break-even point: when the product variety structure changes, changes in direction of the break-even point depends on a variety of products for weight weighted average of the proportion of sales revenue contribution rate changes. When weighted average contribution rate rises, the break-even point will reduce accordingly, on the contrary, when weighted average contribution rate reduce, break-even point will rise accordingly.

6.4　Using CVP to Plan Profit

6.4.1　Realize the Target Profit Analysis

Realizing the target profit analysis is actually an extension of the break-even point analysis. CVP analysis can be used to calculate the volume of sales that would be required to achieve a target level of profit. To achieve a target profit, the business will have to earn enough contribution to cover all of its fixed costs and then make the required amount of profit.

(1) Achieve Pre-tax Profit Model

According to:

Operating profit = Sales quantity × (Sale price - Variable

costs per unit) − Fixed costs
Thus,
Sales unit of achieve target profit
= (Target profit+ Fixed costs)/ Contribution per unit
Sales revenue of achieve target profit
= (Target profit+ Fixed costs)/Contribution rate
Note:
Contribution per unit
= Sale price − Variable costs per unit
Contribution rate
= Contribution / Sales revenue
= 1−Variable costs rate

(2) Achieve After-tax Profit Model
According to:
After-tax profit = Pre-tax profit × (1−Tax rate)
Pre-tax profit = After-tax profit /(1−Tax rate)
Sales unit of achieve target profit
= [After-tax profit / (1−Tax rate) + Fixed costs] / Contribution per unit
Sales revenue of achieve target profit
= [After-tax profit / (1−Tax rate) + Fixed costs] /Contribution rate

Example 6.8

LBO Ltd has capital employed of $1million. Its target return on capital employed is 20% per annum. LBO Ltd manufactures a standard product 「H1N1」.
Selling price of 「H1N1」 = $60 unit
Variable costs per unit = $20
Annual fixed costs = $100,000
Required:
What volume of sales is required to achieve the target profit?

Solution:
Target profit = 1,000,000×20% = $200,000
Target contribution = 100,000+200,000 = $300,000
To achieve the target profit,
Sales (u) = 300,000÷(60−20) = 7,500 unit

6.4.2 Change Related Factors to Affect the Profit of the Target

Based on above formulas, the change of the related factors to affect the target profit can be analysied.

Example 6.9

In example 6.8, if
a) Selling price of 「H1N1」 fell to $50;
b) Variable costs per unit increased to $25;
c) Annual fixed costs increased to $120,000.
Required:
What volume of sales is required to achieve the target profit, respectively?

Solution:
To achieve the target profit,
a) Sales (u) = 300,000÷(50−20) = 10,000
b) Sales (u) = 300,000÷(60−25) = 8,572
a) Sales (u) = (120,000+200,000)÷(60−20)
            = 8,000

6.5 Using CVP for Sensitivity Analysis

Sensitivity analysis means to analyze the effect of uncertain factors on profit. If a factor with little uncertainty can affect the profit, then we can say the factor has strong sensitivity, vice versa. Sensitivity analysis aims to find out the factors to which the profit is most sensible, therefore to provide important information to decision makers.

The sensitivity analysis of the relationship between profit and relevant change factor (volume\price\variable cost per unit\fixed cost) main research on two questions:
One is how changes in the relevant factors will make enterprise by profit into a loss;
Secondly, the relevant factors change for the influence degree of the change on profits.

So the content of the need to master the following two aspects:
(1) The Determination of the Critical Value for the Relevant Factors

本量利分析中影響利潤的相關因素的臨界值是指其他相關因素保持目前水平不變的情況下,利潤達到不盈不虧狀態時某相關因素應有的水平。

The change of sales volume, unit price, variable cost per unit and fixed cost, will have an effect on profit. When this effect is negative and reaches a certain degree, the charge which can make the enterprise profit to zero into the profit and loss critical state is the critical value of the relevant factors. If the change is beyond the level of the critical value, the enterprise will turn into a state of loss. The purpose of sensitivity analysis is to determine the critical value of the relevant factors. In simple terms, the critical value is the minimum allowable values of the primary sales and the unit price and the largest allowable values of variable cost per unit and fixed cost reaching the break-even point.

The critical value of sales quantity
= Fixed cost/ (Price-Variable cost per unit)

The critical value of sale price
= Fixed cost/ Sales quantity + Variable cost per unit

The critical value of variable cost per unit
= Sales price - Fixed cost/Sales quantity

The critical value of fixed cost
= Sales quantity × (Sales price - Variable cost per unit)

(2) The Determination of the Sensitive Coefficient for the Relevant Factors

If some factors which change smaller lead to change greater for profits, these factors is called sensitive factors; If some factors though, which change greater lead to change less for profits, these factors is called the insensitive factors. Corporate decision makers need to know which factor is more sensitive for profits in order to distinguish between primary and secondary factor, and to know that how much degree of the factors change ensure the realization of the target profit. So the sensitive coefficient must be calculated.

若某因素變動導致利潤較大的變動,則該因素為敏感因素;若某因素變動引起利潤的變動較小,則該因素為非敏感因素。相關因素的敏感程度通過計算敏感系數來反應。

Sensitive coefficient
= Percentage changes of profit change/Percentage changes of the factor change

Sensitive coefficient of sales quantity
= (Price-Variable cost per unit) × Sales quantity /Profit

Sensitive coefficient of sale price
= Sale price × Sales quantity /Profit

Sensitive coefficient of variable cost per unit
= Variable cost per unit × Sales quantity/ Profit

Sensitive coefficient of fixed cost
= Fixed cost/Profit

Example 6.10

BM company manufactures a single product GE. In next year, budget sales are 500 units at selling price of $350 per unit. Variable costs remain $230 per unit and fixed costs stay at $35,000.

Required:
a) Calculate the critical value for the relevant factors.
b) Calculate the sensitive coefficient for the relevant factors.

Solution:
a) The critical value for the relevant factors.
The critical value of sales quantity
= 35,000 ÷ (350−230) = 292
The critical value of sale price
= 35,000 ÷ 500 + 230 = 300
The critical value of variable cost per unit
= 350 − 35,000 ÷ 500 = 280
The critical value of fixed cost
= 500×(350 − 230) = 60,000
b) The sensitive coefficient for the relevant factors.
Profit = 500×(350 − 230) − 35,000 = 25,000
Sensitive coefficient of sales quantity
= (350 − 230)×500 ÷ 25,000 = 2.4
Or, if sales quantity increase by 10%,
Profit = 500 × (1+10%) × (350 − 230) − 35,000
= 31,000
Sensitive coefficient of sales quantity
= [(31,000 − 25,000) ÷ 25,000] ÷ 10% = 2.4
Sensitive coefficient of sale price
= 500×350 ÷ 25,000 = 7
Or, if sale price increase by 10%,
Profit = 500 × [350 × (1+10%) − 230] − 35,000
= 42,500
Sensitive coefficient of sale price
= [(42,500−25,000) ÷ 25,000] ÷ 10% = 7
Sensitive coefficient of variable cost per unit
= −230 × 500 ÷ 25,000 = −4.6
Or, if variable cost per unit increase by 10%,
Profit = 500 × [350 − 230 × (1+10%)] − 35,000
= 13,500

Sensitive coefficient of sale price
= [(13,500−25,000)÷25,000]÷10% = −4.6
Sensitive coefficient of fixed cost
= −35,000÷25,000 = −1.4
Or, if fixed cost increase by 10%,
Profit = 500 × (350 − 230) − 35,000×(1+10%)
= 21,500
Sensitive coefficient of sale price
= [(21,500−25,000)÷25,000]÷10% = −1.4

6.6 Effect of Sales Mix on CVP Analysis

Most companies sell more than one product. Selling price and variable costs differ for each product, so each product makes a different contribution to profits. The CVP analysis we used earlier for single product apply to a company with multiple products by weighted average contribution margin of sales mix. Sales mix have two types, according sales quantity or sales revenue, so weighted average contribution margin of sales mix have two formulas, then the CVP analysis for multiple products include following two kinds of situations.

To calculate breakeven sales or target sales in units for each product of sales mix, the following steps should be completed.
Step 1: Identify sales mix according to sales quantity or sales revenue.
Step 2: Calculate the weighted average contribution margin per sales mix.
Step 3: Calculate the breakeven sales or target sales in units of sales mix.
Step 4: Calculate the breakeven sales or target sales in units for each product of sales mix. Multiply the breakeven sales or target sales in units of sales mix by number of each product in a sales mix.

Example 6.11

GM Company manufactures two products P and Q. In the past year, the sales are 8,000 units in the selling price of P product with $50 per unit. Variable of P product is $30 per unit. The sales are 2,000 units in the selling price of Q product with $150 per unit. Variable of Q product is $70 per unit. GM Company's total fixed cost is $192,000.

Required:

a) Calculate the breakeven sales in units for each product of sales mix according sales quantity and according sales revenue.

b) In the next year, GM Company remains the sales structure in the last year. If expected profit is 288,000, calculate the target sales in units for each product of sales mix according sales quantity and according sales revenue.

Solution:

a) Sales mix according sales quantity.

Step 1: In the past year, GM Company sold 8,000 units P product and 2,000 units Q product, so sales mix according to sales quantity is made up of four P products and one Q product.

Step 2: The weighted average contribution margin per sales mix
$$= 4 \times (50-30) + 1\times(150-70) = 160$$

Step 3: The breakeven sales in units of sales mix
$$= 192,000 \div 160 = 1,200$$

Step 4: The breakeven sales in units for each product of sales mix:

Breakeven sales of P product = 1,200×4 = 4,800

Breakeven sales of Q product = 1,200×1 = 1,200

Sales mix according sales revenue:

Step 1: In the past year, GM Company sold 8,000 units P product with $50 per unit and 2,000 units Q product with $150 per unit, the sales revenue is $400,000 P product and $300,000 Q product, so sales mix according sales revenue is made up of four P products and three Q products.

Step 2: The weighted average contribution margin per sales mix
$$= 4\times(50-30)+3\times(150-70) = 320$$

Step 3: The breakeven sales in units of sales mix
$$= 192,000 \div 320 = 600$$

Step 4: The breakeven sales in units for each product of sales mix

Breakeven sales of P product = 600×4 = 2,400

Breakeven sales of Q product = 600×3 = 1,800

b) Sales mix according sales quantity.

Step 1: In the past year, GM Company sold 8,000 units P product and 2,000 units Q product, so sales mix according to sales quantity is made up of four P products and one Q product.

Step 2: The weighted average contribution margin per sales mix

$$= 4\times(50-30) + 1\times(150-70) = 160$$

Step 3: The target sales in units of sales mix

$$= (192,000+288,000) \div 160 = 3,000$$

Step 4: The target sales in units for each product of sales mix

Breakeven sales of P product = 3,000×4 = 12,000
Breakeven sales of Q product = 3,000×1 = 3,000

Sales mix according sales revenue:

Step 1: In the past year, GM Company sold 8,000 units P product with $50 per unit and 2,000 units Q product with $150 per unit, the sales revenue is $400,000 P product and $300,000 Q product, so sales mix according sales revenue is made up of four P products and three Q products.

Step 2: The weighted average contribution margin per sales mix

$$= 4\times(50-30)+3\times(150-70) = 320$$

Step 3: The breakeven sales in units of sales mix

$$= (192,000+288,000) \div 320 = 1,500$$

Step 4: The breakeven sales in units for each product of sales mix

Breakeven sales of P product = 1,500×4 = 6,000
Breakeven sales of Q product = 1,500×3 = 4,500

The CVP analysis can be used in short-term Business Decisions, including special sales order and regular pricing decision, products/departments or territories dropping decision, sell or processing further decision.

QUESTIONS:

Questions 1 and 2 are based on the following data:

Sales units	128,000
Sales revenue	$640,000
Variable costs	$384,000
Fixed costs	$210,000

1. What sales revenue is required to earn a profit of $65,000?
   A. $458,333　　B. $590,000　　C. $687,500　　D. $705,000

2. How many sales units are required to earn a profit of £52,000?
   A. 52,400 units　B. 87,333 units　C. 131,000 units　D. 160,500 units

Questions 3 ~ 5 are based on the following data:
A firm, ACME Cleaning Supply, wishes to sell a single cleaning solution called VeryKleen. The cleaner costs $4.50 per bottle to produce and package. The firm will sell the product using door to door salespeople. A selling price of $9.00 per bottle is expected and the firm anticipates fixed costs of $50,000 per year.

3. What is the contribution margin percentage?
   A. 200%　　B. 50%　　C. $9.00　　D. $4.50

4. What is the breakeven quantity?
   A. 3,023 bottles　B. 11,112 bottles　C. 9,999 bottles　D. 100,008 bottles

5. What is the breakeven revenue?
   A. $1,000,000　B. $11,112　C. $100,008　D. $3,055

The following data relates to questions 6 and 7
Data concerning K Limited's single product is as follows:

	$ per unit
Selling price	6.00
Variable production cost	1.20
Variable selling cost	0.40
Fixed production cost	4.00
Fixed selling cost	0.80

Budgeted production and sales for the year are 10,000 units

6. What is the company's breakeven point, to the nearest whole unit?
   A. 8,000 units　B. 8,333 units　C. 10,909 units　D. 10,910 units

7. It is now expected that the variable production cost per unit and the selling price per unit will each increase by 10%, and fixed production costs will rise by 25%.
   What will be the new breakeven point, to the nearest whole unit?
   A. 8,788 units　B. 11,600 units　C. 11,885 units　D. 12,397 units

8. A company's breakeven point is 6,000 units per annum. The selling price is $90 per unit and the variable cost is $40 per unit. What are the company's annual fixed costs?
   A. $120　　B. $240,000　　C. $300,000　　D. $540,000

9. Z plc makes a single product which it sells for $16 per unit. Fixed costs are $76,800 per month and the product has a contribution/sales ratio of 40%. In a period when actual sales were $224,000, Z plc's margin of safety, in units, was:
A. 2,000　　　　　B. 12,000　　　　　C. 14,000　　　　　D. 32,000

10. Which of the following statements about profit-volume graphs is/are correct?
(i) The profit-volume line crosses the x axis at the breakeven point.
(ii) Any point on the profit-volume line above the x axis indicates the profit (as measured on the vertical axis) at the level of activity.
(iii) The profit-volume line starts at the origin.
A. (i) and (ii) only　　　　　　　　B. (ii) and (iii) only
C. (i) and (iii) only　　　　　　　　D. (i), (ii) and (iii)

11. A company has fixed costs per period as follows:
　　Manufacturing　　　　$56,000
　　Non-manufacturing　　$38,000
Variable costs of the company's single product are $4.20 per unit and the selling price is $7.00 per unit.
What sales revenue (to the nearest $000) is required in a period to make a profit of $6,000?
A. $163,000　　　B. $167,000　　　C. $241,000　　　D. $250,000

12. A product has the following unit costs:
　　Variable manufacturing　　　　$7.60
　　Variable non-manufacturing　　$1.40
　　Fixed manufacturing　　　　　$3.70
　　Fixed non-manufacturing　　　$2.70
The selling price of the product is $17.50 per unit.
What is the contribution/sales ratio?
A. 12.0%　　　　B. 48.6%　　　　C. 51.4%　　　　D. 56.6%

13. 5,400 units of a company's single product were sold for total revenue of $140,400. Fixed costs in the period were $39,420 and net profit was $11,880.
What was the contribution per unit?
A. $7.30　　　　B. $9.50　　　　C. $16.50　　　　D. $18.70

14. A company manufactures and sells 4,000 units of a product each month at a selling price of $22 per unit. The prime cost of the product is $11.60 per unit and the monthly overheads are:

	($)
Variable production	7,200
Variable selling and administration	5,200
Fixed production	16,400
Fixed selling and administration	6,800

What is the product's gross profit margin (to one decimal place)?
A. 6.8%   B. 20.5%   C. 33.2%   D. 59.5%

Questions 15 ~ 16 are based on the following data:
The following planned results are available for a company with a single product:
    Sales units   112,000
    Sales revenue $100,800
    Variable costs $60,480
    Fixed costs $36,000

15. What sales revenue is required to earn a profit of $5,000?
A. $68,333   B. $90,000   C. $102,500   D. $113,889

16. What is the margin of safety (sales units)?
A. 10,800   B. 12,000   C. 22,000   D. 100,000

17. Which of the following describes the margin of safety?
A. actual contribution margin achieved compared with that required to break-even
B. actual sales compared with sales required to break-even
C. actual versus budgeted net profit margin
D. actual versus budgeted sales

18. The following data relates to a company with a single product:
    Selling price $12.50 per unit
    Fixed production costs $77,000 per period
    Fixed non-production costs $46,000 per period
    Break-even sales per period 24,600 units
What is the contribution per unit?
A. $3.13   B. $5.00   C. $7.50   D. $9.37

19. A company has a single product. The following budgeted information relates to a period:
    Sales units   800,000
    Sales revenue $1,000,000
    Total variable costs $590,000
    Total fixed costs $350,000
What sales revenue (to the nearest £'000) is required to break even?
A. $350,000   B. $593,000   C. $683,000   D. $854,000

20. Budgeted sales of a company's single product in a period are 20,000 units, producing a total contribution of $180,000 at a selling price of $24 per unit. Fixed costs are $6 per unit based on the budgeted sales quantity.
What is the budgeted variable cost per unit?
A. $3   B. $9   C. $15   D. $18

21. A firm makes a single product. Budgets have been prepared for the year ahead and include production and sales of 60,000 units with a break-even point of 45,000 units.
What is the margin of safety ratio?
A. 25%　　　　　　B. 33%　　　　　　C. 75%　　　　　　D. 133%

22. A book publisher makes an initial payment of $25,000 to authors for each accepted manuscript, followed by a royalty payment of 15% of the net sales price of each book sold. The net sales price of a book, which is the revenue received by the publisher, is the listed selling price in bookstores less the bookstore margin of 20% of the listed selling price. A particular book has a listed selling price of $15.00. Costs incurred on the book by the publisher (excluding initial and royalty payments to the author) are:
　　Variable costs per copy　　$3.20
　　Total fixed costs　　　　　$80,000
Required:
a) Calculate the number of copies of the particular book that need to be sold for the publisher:
(i) To break even;
(ii) To make a profit of £35,000.
b) Prepare a profit/volume (P/V) chart for the publisher, relating to the particular book publication, covering sales up to 25,000 copies.

23. Company A manufactures and sells a single product. The following information is available:
　　Selling price per unit　$60.00
　　Variable costs per unit　$36.00
　　Fixed costs per period　$216,000
Required:
(i) Draw a profit/volume (P/V) chart based on sales up to 14,000 units per period.
(ii) Clearly identify the break-even point, and areas of profit and loss, on the chart.

24. A garage operates a vehicle repair service. Space is limited and, although the garage is usually busy, the owner is concerned about the amount of profit that can be generated. Summarised data concerning vehicle repairs follows:
　　Average number of repairs per period 85
　　Average variable cost of each repair　$126
　　Average sales value of each repair　$210
The owner is considering extending the garage opening hours. This would result in an increase in fixed costs from $6,972 to $7,728 per period. The average variable cost and the average sales value of each repair would be expected to remain the same.
Required:
a) For the current situation, calculate per period the:
(i) profit.
(ii) break-even sales revenue.
b) For the proposed extended opening hours, calculate the average number of repairs required per period to achieve the current level of profit.

25. The variable costs per unit of a company's single product for the period just ended were:

	$
Production	120
Non-production	16

The selling price of the product in the period was $200 per unit and the sales revenue required to break-even was $120,000.

Required:
a) Calculate for the period just ended:
(i) The contribution/sales ratio.
(ii) The total fixed costs.
b) In the following period it is expected that fixed costs will total $39,000.

Required:
Calculate the required contribution per unit in the following period for the break-even point to be 500 units.

26. The total costs incurred at various output levels, for a process operation in a factory, have been measured as follows:

Output (units)	Total cost ($)
11,500	102,476
12,000	104,730
12,500	106,263
13,000	108,021
13,500	110,727
14,000	113,201

a) Using the high-low method, analyse the costs of the process operation into fixed and variable components.
b) Calculate, and comment upon, the break-even output level of the process operation in above, based upon the fixed and variable costs identified and assuming a selling price of $10.60 per unit.
c) Calculate the target output level if the company wishes to make a profit of $30,000.
(i) Define the term「margin of safety」.
(ii) Calculate the margin of safety if budgeted output is 12,500 units.

27. A firm makes a single product with a marginal cost of $0.15. Up to 10,000 units can be sold at $0.40 per unit but when the selling price is reduced to $0.30 per unit, total units sold will be more than 10,000. Fixed costs are $2,500 per period and there is a planned profit of $4,000 per period.
a) Based on selling price of $0.30, how many units must be made and sold to break even?
b) Based on selling price of $0.20, how many units must be made and sold to break even?
c) How many units must be made and sold to achieve profit of $4,000 per period?
d) What is the margin of safety for the period?

Chapter 7 Budgeting

Learning Objectives

After the study of this chapter, you should be able to:
a) Explain why organizations use budgeting and the benefits of budgeting.
b) Describe the planning and control cycle in an organization.
c) Prepare the operating budgets (sale, production, ending inventory, direct materials, direct labor, manufacturing overheads, selling and administrative budgets).
d) Prepare financial budgets (cash budgets, budget income statement and budget balance statement).
e) Explain sensitivity analysis, variance analysis and rolling up unit budgets.

7.1 Budgeting Overview

Before this chapter, we have focused on how costs are created by short-term and long-term decisions. We called costs that varied with the activity level in the company (variable costs); Costs that did not change with changes in activity levels, we called fixed costs. For decision making in the short-term, the company's fixed costs are considered to be given and fixed. So costs relevant to the company in the short run are variable costs, since they are the only ones that are controllable.

In this chapter, we are going to discuss the budgeting process, which determines the planned level of most variable costs. In addition, you will see throughout this chapter, knowing how costs behave continues to be important when forming budgets. Total fixed costs will not change as sales changes within the relevant range. However, total variable costs must be adjusted when sales is expected to fluctuate.

7.1.1 Using Budgets to Plan and Control

Most families have developed a financial plan to guide them in allocating their resources over a planning period. Usually the plan reflects spending priorities and demands, including specific spending categories such as the food, clothing, mortgage, utilities and so on. Family budgets usually are the result of discussions and negotiations among parents and children reflecting their different priorities and needs. For example, money left over after required spending may go into savings or be used for other purpose; One parent may want to use most of the disposable income for a vacation, while another may want to use the money to buy a new car. Within the same household, children may ask their parents for new toys or new bike. The family budgets is a planning tool, but it also serves as a control on the behavior of family members by setting limits on what can be spent in each budget category. Without a budget, the family would not have a way to monitor and control its spending by categories of spending. Without such monitoring and control, a family can easily succumb to unexpected debt and severe financial difficulties.

Budgets serve the same purpose for managers within the business units of an organization and are central part of the design and operation of management accounting systems. Large for-profit companies, such as Alibaba Group, and non-profit organizations, such as Red Cross, use budgets to plan and control actions and the related revenues and expenses. Figure 7.1 shows the central role budgets play the relationship between planning and control.

Fugure 7.1　Relationship Between Budget, Planning and Control

As in household, budgets in organizations reflect in quantitative terms how to allocate financial resources to each part of an organization department or division or other distinct part based on planned activities and short run objectives of that part of organization. For example, a bank manager may want to increase the bank's local market share, which may require a larger spending budget than the previous year's for local advertising, implementing a staff training program to improve customer service, and renovating the building to make it more appealing to customers.

7.1.2 Benefits of Budgeting

There are three key benefits of budgeting. Budgeting forces managers to plan, promotes coordination and communication, and provides a benchmark for evaluating actual performance.

預算的作用：計劃、協調與溝通、為業績評價提供依據。

Please always keep in mind: A budget is a quantitative expression of the planned, money inflows and outflows that reveal whether the current operating or business plan will meet the organization's financial objectives. Budgeting is the process of preparing budgets.

Budget：業務營運過程中資金流入和流出的量化表達。

Budgeting：編制預算。

Budgets also provide a way to communicate the organization's short-term goals to its employees. Asking organization unit managers to undertake budgeting activities can accomplish two things:

預算提供一種溝通的方式，將企業的短期目標傳達給員工。

a) Reflect how well unit managers understand the organization's goals, so that they can align their activities and spending priorities with those goals.

b) Provide an opportunity for the organization's senior planners to correct misconceptions about the organization's goals. Suppose an organization recognized quality as a critical factor for its success and wanted to promote quality awareness among its employees. If a department prepared a budget that reflected no expenditures for employee quality training, a senior planner would recognize that the organization's goal with respect to quality had not been communicated properly to the person who should have recognized the need for quality training.

編制預算有助於對各種經營活動進行協調。

Budgeting also serves to coordinate many activities of an organization. For example, budgets show the effect of sales levels on purchasing, production, and administrative activities and the number of employees that must be hired to serve customers. In this sense, budgeting is a tool that promotes coordination of a organization's activities and helps identify coordination problems. Suppose the sales force plans to significantly expand sales. By comparing selling plans with manufacturing capacity, planners might discover that the manufacturing operations are unable to meet increasing sales level of the plan. The kind of coordination needed can be accomplished through powerful desktop computers and software; With the help computer and software, planners can simulate the effect of different decisions on the organization's financial, human, and physical resources. Simulation analysis—which is simply, what-if analysis—helps managers choose a course of action among many alternatives by identifying a decision's consequences in a complex system with inter-dependencies.

By considering the interrelationships among operating activities, a budget helps to anticipate potential problems and serve as a tool to help provide solutions to these problems. For example, Ice cream factory engage in seasonal production, consuming large amounts of cash when they build inventory during the ice cream season. Throughout the year, the ice cream sells its inventory and recovers cash. Budgeting reflects this cash cycle, shown in figure 7.2, and provides information to help the organization plan the borrowing needed to finance the inventory buildup early in the cash cycle. If budget planning suggests that the organization's sales potential exceeds its manufacturing potential, the organization can develop a plan to put more capacity in place or to reduce planned sales. It is important for managers to anticipate problems because putting new capacity in place can take several months to several years.

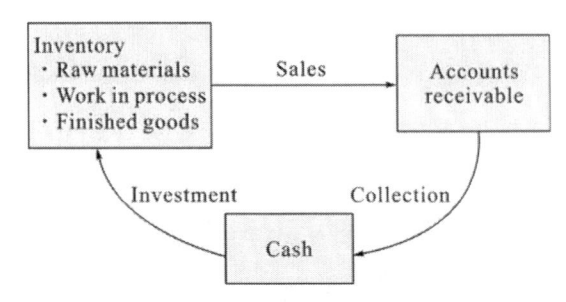

Figure 7.2　Cash Cycle

7.1.3 Elements of Budgeting

编制预算:预测不同时期对四种资源的需求。

Budgeting involves forecasting the demand for four types of resources over different time periods:

a) Flexible resources that create variable costs. Flexible resources are those that can be acquired or disposed of in the short term, such as the plastic, glue, and varnish used in a pencil factory or, based on estimates of the number of automobiles to be assembled, the number of tires an automobile assembly plant needs to acquire.

b) Resources that, in the intermediate run and long run, enhance the potential of the organization's strategy. These are discretionary expenditures, which include research and development, employee training, the maintenance of capacity resources, advertising, and promotion. These discretionary expenditures do not provide capacity, nor do they vary with the level of organizational activity.

c) Intermediate-term capacity resources that create fixed costs. An example is forecasting the need for equipment's' insurance that might be contracted on a quarterly, semiannual, or annual basis.

d) Long-term capacity resources that create fixed costs. For example, a new production facility for a medical devices manufacturer, which might take several years to plan and build and might be used for 10 years.

7.2 The Components of the Master Budgeting

全面预算包括业务预算和财务预算。

The framework for budgeting in organizations is discussed in the following sections. The discussion begins with the budgeting process and leads to formulation of the master budget. Two major types of budgets make up the master budget:

a) Operating budgets summarize the level of activities such as sales, purchasing, and production.

b) Financial budgets, such as balance sheets, income statements, and cash flow statements, identify the expected financial consequences of the activities summarized in the operating budgets.

The master budget is the set of budgeted financial statements and supporting schedules for the entire organization. Figure 7.3 shows the order in which managers prepare the component of the master budget for merchandise such as Taobao.

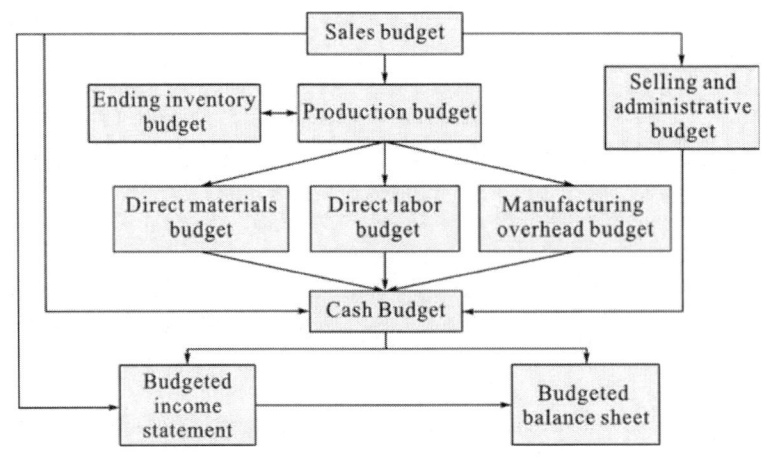

Figure 7.3　Component of Master Budget

Figure 7.3 summarizes different components of the budget. The dashed lines from the cash budget show how the estimated financial consequences from the organization's tentative budgets can influence the organization's plans and objectives. The dashed lines illustrate a recursive process in which planners compare projected financial results with the organization's financial goals. If initial budgets proved infeasible (because the organization does not have the capacity to produce or sell the planned level of output) or financially unacceptable (because the proposal plan does not yield the desired target level of profits), planners will repeat the budgeting cycle with a new set of decisions until the results are both feasible and financially acceptable.

The master budget process in figure 7.3 includes two broad sets of outputs:

a) The operating budgets that operating personnel use to guide operations are sales budget, selling and administrative budget, and production budget (including ending inventory budget, direct materials budget, direct labor budget and manufacturing overhead budget).

b) The expected or budgeted financial results. Planners usually present the expected, or budgeted, financial results, in three forms:

(i) A statement of expected cash flows.
(ii) The budgeted balance sheet.
(iii) The budgeted income statement.

7.3 Preparing the Operating Budget

業務預算包括：各產品銷售預算、生產預算、產成品庫存預算、直接材料預算、直接人工預算、生產(製造)費用預算、非生產費用(銷售和管理費用)預算。

Operating budgets typically consist of the following seven budgets (look at figure 7.3):

a) The sales budget identifies the planned level of sales for each product.
b) The production budget schedules required production.
c) The ending inventory budget calculates the cost of the finished goods inventory at the end of each budget period.
d) The direct materials budget specifies the total amount of all kinds of materials that directly used for produce products.
e) The direct labor budget specifies the total amount of labor costs that directly used for produce products.
f) The manufacturing overhead budget indicates all of the costs that a factory incurs, other than the variable costs required to build products, such as direct materials and direct labor within a reporting period.
g) The selling and administrative budget includes administration, staffing, research and development, and advertising.

Operating budgets specify the expected resource requirements of selling, capital spending, manufacturing, purchasing, labor management, and administrative activities during the budget period. Operations personnel use those plans represented in the operating budget to guide and coordinate the level of various activities during the budget period. At the same time, operations personnel record data from current operations that can be used to develop future budgets.

7.3.1 The Sale Budget

銷售預算：各產品的預計銷售量和銷售額。

The forecast of sales revenue is the cornerstone of the master budget because the level of sales affects expenses and almost all other elements of the master budget. The sales budget contains an itemization of a company's sales expectations for the budget period, in both units and dollars. If a company has a large number of products, it usually aggregates its expected sales into a smaller number of product categories or geographic regions; Otherwise, it becomes too difficult to generate sales estimates for this budget. The sales budget is usually presented in either a monthly or quarterly format; Presenting only annual sales information is too aggregated, and so provides little actionable information.

The information in the sales budget comes from a variety of sources. Most of the detail for existing products comes from those personnel who deal with them on a day-to-day basis. The marketing manager contributes sales promotion information, which can alter the timing and amount of sales. The engineering and marketing managers may also contribute information about the introduction date of new products, as well as the retirement date of old products. The chief executive officer may revise these figures for the sales of any subsidiaries or product lines that the company plans to terminate or sell during the budget period.

It is generally best not to include in the sales budget any estimates for sales related to prospective acquisitions of other companies, since the timing and amounts of these sales are too difficult to estimate. Instead, revise the sales budget after an acquisition has been finalized.

The basic calculation in the sales budget is to itemize the number of unit sales expected in one row, and then list the average expected unit price in the next row, with the total revenues appearing in a third row. The unit price may be adjusted for marketing promotions. If any sales discounts or returns are anticipated, these items are also listed in the sales budget.

因為銷售預算的訊息貫穿全面預算,涉及各預算,所以準確預測銷售很重要。

It is extremely important to do the best possible job of forecasting, since the information in the sales budget is used by most of the other budgets (such as the production budget and the direct materials budget). Thus, if the sales budget is inaccurate, then other budgets that use it as a source material will be influenced as well.

It is quite difficult to derive a sales forecast that proves to be accurate for any period of time, so an alternative is to periodically adjust the sales budget with revised estimates, perhaps on a quarterly basis. If this is done, the rest of the budget that is derived from the sales figures will also have to be revised, which require a significant amount of staff time. The projected unit sales information in the sales budget feeds directly into the production budget, from which the direct materials and direct labor budgets are created. The sales budget is also used to give managers a general sense of the scale of operations, when they create the overhead budget, sales and administrative expenses budget. The total net sales dollars listed in the sales budget are carried forward into the revenue line item in the master budget.

Figure 7.4 is an example of the Sales Budget: ABC Company plans to produce an array of plastic balls during the upcoming budget year, all of which fall into a single product category. Its sales forecast is outlined as follows:

ABC Company
Sales Budget
For the Year Ended December 31, 20XX

	Quarter 1	Quarter 2	Quarter 3	Quarter 4
Forecasted unit sales	5,500	6,000	7,000	8,000
x Price per unit	$10	$10	$11	$11
Total gross sales	$55,000	$60,000	$77,000	$88,000
- Sales discounts & allowances	$1,100	$1,200	$1,540	$1,760
= Total net sales	$53,900	$58,800	$75,460	$86,240

Figure 7.4 An Example of the Sales Budget

ABC's sales manager expects that increased demand in the second half of the year will allow it to increase its unit price from $10 to $11. Also, the sales manager expects that the company's historical sales discounts and allowances percentage of two percent of gross sales will continue through the budget period.

This example of the sales budget is simplistic, since it assumes that the company only sells in one product category. In reality, this example might have been a detail page that rolls up into the main sales budget, where it would occupy a single line item.

7.3.2 The Production Budget

生產預算:計算在預算期內的產量。

The production budget calculates the number of units of products that must be manufactured, and is derived from a combination of the sales forecast and the planned amount of finished goods inventory to have on hand (usually as safety stock to cover for unexpected increases in demand). The production budget is typically prepared for a 「push」 manufacturing system, as is used in a material requirements planning environment.

The production budget is typically presented in either a monthly or quarterly format. The basic calculation used by the production budget is:

+ Forecasted unit sales
+ Planned finished goods ending inventory balance
= Total production required
− Beginning finished goods inventory
= Products to be manufactured

It can be very difficult to create a comprehensive production budget that incorporates a forecast for every variation on a product that a company sells, so it is customary to aggregate the forecast information into broad categories of products that have similar characteristics.

The planned amount of ending finished goods inventory can be subject to a considerable amount of debate, since having too much may lead to obsolete inventory that must be disposed of at a loss, while having too little inventory can result in lost sales when customers want immediate delivery. Unless a company is planning to draw down its inventory quantities and terminate a product, there is generally a need for some ending finished goods inventory.

Figure 7.5 is an example of a production budget, ABC Company plans to produce an array of plastic pails during the upcoming budget year, all of which fall into the general product a category. Its production needs are outlined as follows:

ABC Company
Production Budget
For the Year Ended December 31, 20XX

	Quarter 1	Quarter 2	Quarter 3	Quarter 4
Forecasted unit sales	5,500	6,000	7,000	8,000
+ Planned ending inv. units	500	500	500	500
= Total production required	6,000	6,500	7,500	8,500
- Beginning F/G inventory	1,000	500	500	500
= Units to be manufactured	5,000	6,000	7,000	8,000

Figure 7.5　An Example of a Production Budget

The planned ending finished goods inventory at the end of each quarter declines from an initial 1,000 units to 500 units, since the materials manager believes that the company is maintaining too much finished goods in stock. Consequently, the plan calls for a decline from 1,000 units of ending finished goods inventory at the end of the first quarter to 500 units by the end of the second quarter, despite a projection for rising sales. This may be a risky forecast, since the amount of safety stock on hand is being cut while production volume increases by over 30 percent. Given the size of the projected inventory decline, there is a fair chance that ABC will be forced to increase the amount of ending finished goods inventory later in the year.

The production budget deals entirely with unit volumes. Unlike most other parts of the corporate budget, the production budget does not translate its production requirements into dollars. Instead, the unit requirements of the production budget are shifted into other parts of the budget, such as the direct labor budget and the direct materials budget, which are then translated into dollars.

A case can be made that this budget is not needed in a 「pull」 production environment, where goods are produced only on an as-needed basis. Under this concept, it is not necessary to estimate unit quantities to be produced, since the production environment merely reacts to actual demand.

7.3.3 The Ending Inventory Budget

產成品庫存預算:計算在預算期末,庫存產成品的成本。

The ending Inventory (ending finished goods inventory) budget calculates the cost of the finished goods inventory at the end of each budget period. It also includes the unit quantity of finished goods at the end of each budget period, but the real source of that information is the production budget.

The primary purpose of this budget is to provide the amount of the inventory asset that appears in the budgeted balance sheet, which is then used to determine the amount of cash needed to invest in assets. If you do not intend to create a budgeted balance sheet, there is no need to create an ending finished goods inventory budget. When a company needs to closely monitor its cash balances on an ongoing basis, the ending finished goods inventory budget should not only be created, but also updated on a regular basis.

The ending finished goods inventory budget contains an itemization of the three main costs that are required to be included in the inventory asset under both generally accepted accounting principles and international financial reporting standards. These costs and their derivation are:

a) Direct materials. The cost of materials per unit (as listed in the direct materials budget), multiplied by the number of ending units in inventory (as listed in the production budget).

b) Direct labor. The direct labor cost per unit (as listed in the direct labor budget), multiplied by the number of ending units in inventory (as listed in the production budget).

c) Overhead allocation. The amount of overhead cost per unit (as listed in the manufacturing overhead budget), multiplied by the number of ending units in inventory (as listed in the production budget).

If there are many types of products expected to be in ending inventory, it may be too difficult to calculate this budget on an item-by-item basis. If so, an alternative is to create an approximate cost per unit based on general classifications of inventory types; This derivation is usually based on historical costs, modified for costs expected to be incurred during the budget period.

Figure 7.6 is an example of ending inventory budget. Georgia Corporation sells a single product, and has derived its main cost components in the product budget, direct materials budget, and manufacturing overhead budget. Its ending finished goods inventory cost calculation follows:

Georgia Corporation
Ending Finished Goods Inventory Budget
For the Year Ended December 31, 20XX

	Qtr 1	Qtr 2	Qtr 3	Qtr 4
Cost per unit:				
+ Direct materials cost	$12.50	$12.50	$12.75	$12.75
+ Direct labor cost	4.00	4.50	4.50	4.50
+ Manufacturing O/H cost	6.50	6.50	6.50	6.75
= Total cost per unit	$23.00	$23.50	$23.75	$24.00
Ending finished goods units	8,000	12,000	10,000	9,000
x Total cost per unit	$23.00	$23.50	$23.75	$24.00
= Ending F/G inventory	$184,000	$282,000	$237,500	$216,000

Figure 7.6　An Example of Ending Inventory Budget

Georgia Corporation is expecting the cost of materials to increase in the third quarter, as well as to boost its direct labor cost in the second quarter as the result of a union agreement, and to increase its manufacturing overhead cost in the fourth quarter because of a scheduled rent increase. All of these factors increase the cost of the product from $23.00 per unit at the end of the first quarter to $24.00 at the end of the fourth quarter.

7.3.4　The Direct Material Budget

直接材料預算：計算在預算期內，為滿足生產需求及庫存採購的原料。

The direct materials budget calculates the materials that must be purchased, by time period, in order to fulfill the requirements of the production budget. It is typically presented in either a monthly or quarterly format in the annual budget. In a business that sells products, this budget may contain a majority of all costs incurred by the company, it should be compiled with considerable care. Otherwise, the result may erroneously indicate excessively high or low cash requirements to fund materials purchases.

The basic calculation used by the direct materials budget is:

+ Raw materials required for production
+ Planned ending inventory balance
= Total raw materials required
− Beginning raw materials inventory
= Raw materials to be purchased

It is impossible to calculate the direct materials budget for every component in inventory, since the calculation would be massive. Instead, it is customary to either calculate the approximate amount of inventory required, expressed as a grand total for the entire inventory, or else at a somewhat more detailed level by commodity type. It is possible to create a reasonably accurate direct materials budget by either means, if you have a material requirements planning software package that has a planning module. By entering the production budget into the planning module, the software can generate the expected direct materials budget for future periods. Otherwise, you will have to calculate the budget manually.

A lesser alternative is to calculate the direct materials budget based on the historical percentage of direct materials experienced in recent reporting periods; Doing so assumes that the same ratio of direct material costs to revenues will continue, which can be a dangerous assumption. Realistically, the mix of products sold will change over time, so the historical percentage of direct materials to revenues may not match actual results in future periods.

The preparation of the direct materials budget can be so detailed that the preparer becomes lost in the details and does not determine whether the entire result is reasonable. Accordingly, be sure to review the completed budget based on historical percentages, and consult with the purchasing staff to see if cost assumptions are reasonable.

If product life cycles are quite short and margins vary substantially by product, this budget may become highly inaccurate if the forecast period is for a full year. In this case, it may make more sense to budget over a shorter period.

It is not customary to include a cash requirements calculation as part of the direct materials budget. Instead, the cash requirements are calculated for all of the revenues and expenditures of a business as a whole, and are then summarized on a separate page of the budget.

7.3.5 The Direct Labor Budget

直接人工預算:計算預算期內生產各項產品所需的人工小時,更詳細的預算還區分不同工種所需人工小時。

The direct labor budget is used to calculate the number of labor hours that will be needed to produce the units itemized in the production budget. A more complex direct labor budget will calculate not only the total number of hours needed, but also break down this information by labor category. The direct labor budget is useful for anticipating the number of employees who will be needed to staff the manufacturing area throughout the budget period. This allows management to anticipate hiring needs, as well as when to schedule overtime, and when layoffs are likely. The budget provides information at an aggregate level, but is not typically used for specific hiring and lay off requirements.

The direct labor budget is typically presented in either a monthly or quarterly format. The basic calculation used by the direct labor budget is to import the number of units of production from the production budget, and to multiply this by the standard number of labor hours for each unit. This yields a subtotal of the direct labor hours needed to meet the production target. You can also add more hours to account for production inefficiencies, which increases the amount of direct labor hours. Then multiply the total number of direct labor hours by the fully burdened direct labor cost per hour, to arrive at the total cost of direct labor.

If you have a material requirements planning software package that has a planning module, you may be able to load the production budget into the planning module and have it calculate the required number of direct labor hours, by position. Otherwise, you will have to calculate this budget manually.

It is not customary to include a cash requirements calculation as part of the direct labor budget. Instead, the cash requirements are calculated for all of the revenues and expenditures of a business as a whole, and are then summarized on a separate page of the budget.

You may find that it is too time-consuming to create a labor budget in detail when there are a multitude of labor classifications, since it is extremely difficult to match the budgeted pay levels to real world staffing. Instead, ongoing turnover in all of the pay classifications will inevitably result in mismatches between what the budget says the company should be paying and what it is actually paying for labor.

Another issue is that direct labor may be an essentially fixed cost within a broad range of production volumes, in which case a detailed roll-up of the cost will not necessarily generate better information. If so, it may be sufficient to estimate the direct labor cost based on historical results, and adjusted for expected changes in labor rates.

7.3.6 The Manufacturing Overhead Budget

The manufacturing overhead budget contains all manufacturing costs other than the costs of direct materials and direct labor (which are itemized separately in the direct materials budget and the direct labor budget). The information in the manufacturing overhead budget becomes part of the cost of goods sold of line item in the master budget.

Also, the total of all costs in this overhead budget are converted into a per-unit overhead allocation, which is used to derive the cost of ending finished goods inventory, and which in turn is listed on the budgeted balance sheet. The information in this budget is among the most important of the various departmental budget models, since it may contain a large proportion of the total amount of a company's expenditures. This budget is typically presented in either a monthly or quarterly format.

A less-common format for the overhead budget is to group the line items into fixed and variable expense classifications. It can be difficult to determine the fixed or variable status of a cost, in which case you can add a third cost grouping for mixed costs that contain both fixed and variable cost characteristics. Separate treatment of variable expenses is useful if you want to create a flexible budget, where the budgeted amount of variable costs change to match the amount of actual revenues earned.

In a simplified budgeting environment, the overhead budget may be as simple as an overhead rate that is multiplied by some form of activity, such as direct labor hours or machine time used. This approach is generally not recommended, since it does not reveal the precise nature of the various types of expenses incorporated into the overhead rate, and could even be used by a less ethical manager to increase his manufacturing budget without making it visible to the rest of the management team.

Given the considerable size of the expenditures in this budget, one must guard against the inclusion of an incorrect figure, since the result could be a seriously incorrect overall budget. One way to spot incorrect figures is to match the budgeted totals by period against the actual amounts in-

生產(製造)費用預算決定了成本費用的分配,決定了產成品庫存成本。

curred for the same periods in the immediately preceding year, for reasonableness.

It is not customary to include a cash requirements calculation as part of the manufacturing overhead budget. Instead, the cash requirements are calculated for all of the revenues and expenditures of a business as a whole, and are then summarized on a separate page of the budget.

Figure 7.7 is an example of the manufacturing overhead budget. Delphi Furniture produces Greek-style furniture. It budgets the wood raw materials and cost of its artisans in the direct materials budget and direct labor budget, respectively. Its manufacturing overhead costs are outlined as follows:

Delphi Furniture
Manufacturing Overhead Budget
For the Year Ended December 31, 20XX

	Quarter 1	Quarter 2	Quarter 3	Quarter 4
Administrative salaries	$142,000	$143,000	$144,000	$145,000
Administrative payroll taxes	10,000	10,000	11,000	11,000
Depreciation	27,000	27,000	29,000	29,000
Freight in and out	8,000	7,000	10,000	9,000
Rent	32,000	32,000	32,000	34,000
Supplies	6,000	5,000	7,000	6,000
Travel and entertainment	3,000	3,000	3,000	3,000
Utilities	10,000	10,000	10,000	12,000
Total manufacturing overhead	$238,000	$237,000	$236,000	$237,000

Figure 7.7　An Example of Manufacturing Overhead Budget

The administrative salaries line item contains the wages paid to manufacturing supervisors, the purchasing staff, production clerks, and logistics planning staff, and gradually increases over time to reflect changes in pay rates. The depreciation expense is relatively fixed, though there is an increase in the third quarter that reflects the purchase of new equipment. Both the freight and supplies expenses are closely linked to actual production volume, and so their amounts fluctuate in conjunction with planned production levels. The rent expense is a fixed cost, which increase in the fourth quarter reflecting an increase in scheduled rent.

The budget could also include a calculation of the overhead rate. For example, direct labor hours could be included at the bottom of the budget, which are divided into the total manufacturing overhead cost per quarter to arrive at the allocation rate per direct labor hour.

7.3.7 The Selling and Administrative Budget

所有非生產部門的費用,如銷售部、財務部、技術部、設備部門等。

可採用作業成本法,分析當銷售或資本支出變化時需要哪項作業、多少作業,從而確定該費用。

Incremental budgeting 直譯為增量預算。

Much of the information in this budget can be estimated from historical results, if the types of products manufactured and production volumes do not vary significantly from prior periods.

The selling and administrative expense budget is comprised of the budgets of all non-manufacturing departments, such as the sales, marketing, accounting, engineering, and facilities departments. In aggregate, this budget can rival the size of the production budget, and so is worthy of considerable attention. The selling and administrative expense budget is typically presented in either a monthly or quarterly format. It may also be split up into segments for separate sales and marketing budget and a separate administration budget. The information in this budget is not directly derived from any other budgets. Instead, managers use the general level of corporate activity to determine the appropriate level of expenditure. This can involve activity-based costing analysis to determine which activities are likely to be needed more or less as sales levels and capital spending change. There may also be some impact of bottleneck operations on the amount of expenditures in this budget (especially if the bottleneck is in the sales department). When creating this budget, it is useful to determine the activity levels at which step costs may be incurred, and to incorporate them into the budget.

It is very common to derive the amounts in the sales and administrative expense budget with incremental budgeting, which means that the amounts budgeted are based on the most recent budget or the most recent actual results. This is not the best way to create budgets, since it tends to perpetuate existing spending patterns, and allows managers to retain excess funding. However, since it is a simple way to create a budget, it is the most common method for doing so, especially in companies that are not under significant competitive pressure to cut costs.

It is not customary to include a cash requirements calculation as part of this budget. Instead, the cash requirements are calculated for all of the revenues and expenditures of a business as a whole, and are then summarized on a separate page of the budget.

Figure 7.8 is an example of the selling and administrative budget. ABC Company has sales, marketing, accounting, and corporate employees, as well as related support functions. It creates the following budget for them:

Selling and Administrative Expense Budget
For the Year Ended December 31, 20XX

	Quarter 1	Quarter 2	Quarter 3	Quarter 4
Advertising	$5,000	$5,000	$10,000	$10,000
Insurance	2,000	2,000	2,000	2,000
Payroll taxes	1,500	1,600	1,700	1,800
Rent	8,000	8,000	8,000	8,000
Salaries	20,000	21,000	22,000	23,000
Supplies	1,500	1,500	1,500	1,500
Travel and entertainment	2,500	2,500	2,500	2,500
Utilities	3,000	3,000	3,000	3,000
Other expenses	1,500	1,500	1,500	1,500
Total Expenses	$45,000	$46,100	$52,200	$53,300

Figure 7.8　An Example of the Selling Administrative Budget

The preceding example reveals a common characteristic of most line items in a sales and administrative expense budget, which is that the majority of costs are fixed in the short term, and so do not vary from quarter to quarter. In the example, there is a scheduled increase in the advertising expense in the third quarter, and there are budgeted pay increases in all periods. Otherwise, there are no expectations for cost changes in the budget, and that is a realistic expectation for many companies.

7.4　Preparing the Financial Budget

財務報表預算編制具體包括編制現金預算、編制利潤表預算、編制資產負債表預算。

Planners prepare the budgeted balance sheet and budgeted income statement to estimate the financial consequences of investment, production, and sales budgets. Planners use the statement of budgeted cash flows in two ways:
a) To plan when excess cash will be generated so that it can be used to make short-term investments rather than simply holding cash during the short term.
b) To plan how to meet any cash shortages.

7.4.1　The Cash Budget

The cash budget contains an itemization of the projected sources and uses of cash in a future period. This budget is used to ascertain whether company operations and other activities will provide a sufficient amount of cash to meet projected cash requirements. If not, management must find additional funding sources.

The inputs to the cash budget come from several other budgets. The results of the cash budget are used in the financing budget, which itemizes investments, debt, and both interest income and interest expense.

The cash budget is comprised of two main areas, which are sources of cash and uses of cash. The sources of cash section contains the beginning cash balance, as well as cash receipts from cash sales, accounts receivable collections, and the sale of assets. The uses of cash section contains all planned cash expenditures, which comes from the direct materials budget, direct labor budget, manufacturing overhead budget, selling and administrative expense budget. It may also contain line items for fixed asset purchases and dividends to shareholders.

If there are any unusually large cash balances indicated in the cash budget, these balances are dealt with in the financing budget, where suitable investments are indicated for them. Similarly, if there are any negative balances in the cash budget, the financing budget indicates the timing and amount of any debt or equity needed to offset these balances.

如果現金預算出現異常盈餘(赤字)，則由投資(融資)預算應對。

Cash balances may fluctuate considerably within a single accounting period, thereby masking cash shortfalls that can put a company in serious jeopardy. To spot these issues, it is quite common to create and maintain cash forecasts on a weekly basis. Though these short-term budgets are reasonably accurate for perhaps a month, the precision of forecasting declines rapidly thereafter, so many companies then switch to budgeting on a monthly basis. In essence, a weekly cash budget begins to lose its relevance after one month, then is largely inaccurate after two months.

Figure 7.9 is an example of the cash budget, showing the sources and uses of cash by week:

Everson Manufacturing
Cash Budget

	Week 1	Week 2	Week 3	Week 4
Beginning cash	$25,000	$55,000	-$24,000	-$63,000
Sources of Cash				
+ Cash sales	+10,000	+12,000	+15,000	+18,000
+ Accounts receivable collected	+180,000	+185,000	+180,000	+192,000
+ Asset sales	+30,000	0	+10,000	+25,000
= Total cash available	$245,000	$252,000	$181,000	$172,000
Uses of Cash				
- Direct materials	-$87,000	-$91,000	-$99,000	-$107,000
- Direct labor	-19,000	-20,000	-23,000	-25,000
- Manufacturing overhead	-29,000	-30,000	-34,000	-37,000
- Selling & administrative	-35,000	-35,000	-38,000	-38,000
- Asset purchases	-20,000	0	-50,000	0
- Dividend payments	0	-100,000	0	0
= Total uses of cash	-$190,000	-$276,000	-$244,000	-$207,000
Net Cash Position	$55,000	-$24,000	-$63,000	-$35,000

Figure 7.9　Everson Manufacturing Cash Budget

The example shows that an inordinately large dividend payment in the second week of the cash budget, coupled with a large asset purchase in the following week, places the company in a negative cash position. Paying out such a large dividend can be a problem for lenders, who do not like to issue loans so that companies can use the funds to pay their shareholders and thereby weaken their ability to pay back the loans. Thus, it may be wiser for the company to consider a small dividend payment and avoid a negative cash position.

7.4.2　The Budget Income Statement

The budgeted income statement contains all of the line items found in a normal income statement, except that it is a projection of what the income statement will look like during future budget periods. It is compiled from a number of other budgets, the accuracy of which may vary based on the realism of the inputs to the budget model.

The budgeted income statement is extremely useful for testing whether the projected financial results of a company appear to be reasonable. When used in combination with the budgeted balance sheet, it also reveals scenarios that are not financially supportable (such as requiring large amounts of debt), which management can remedy by altering the underlying budget assumptions.

The budgeted income statement works best when presented for all of the budget periods at once, so that you can compare the results for the various periods and spot anomalies that may require additional investigation.

For analysis purposes, the number of line items in a budgeted income statement may be compressed or contracted in comparison to the line items normally used for an actual income statement. Ideally, the line items should be similar, since the budgeted results are typically loaded into the accounting software for each actual income statement line item and then used in budget-versus-actual reports.

Figure 7.10 is an example of budgeted income statement.

Very Large Corporation
Budgeted Income Statement
For the Year Ended December 31, 20XX

Line Item	Source Budget	Amount
Net sales	Sales budget	$10,000,000
Less: cost of goods sold	(Cost in the ending F/G* inventory budget) x (Sales budget units)	6,500,000
Gross margin		3,500,000
Less: Selling & admin. expenses	Selling and admin. expense budget	3,250,000
Net operating income		250,000
Less: interest expense	Financing budget	75,000
Net income		$175,000

* F/G = Finished goods

Figure 7.10　An Example of Budgeted Income Statement

7.4.3　The Budget Balance Sheet

The budgeted balance sheet contains all of the line items found in a normal balance sheet, except that it is a projection of what the balance sheet will look like during future budget periods. It is compiled from a number of supporting calculations, the accuracy of which may vary based on the realism of the inputs to the budget model.

The budgeted balance sheet is extremely useful for testing whether the projected financial position of a company appears to be reasonable. It also reveals scenarios that are not financially supportable (such as requiring large amounts of debt), which management can remedy by altering the underlying budget model.

A budgeted balance sheet should be constructed in each period spanned by the budget model, rather than ending period only, so that the budget analyst can determine whether the cash flows estimated to be generated will be sufficient to provide adequate funding for the company throughout the budget period.

The totals of the asset side are supposed to equal to the liability and equity side in the balance sheet, but this may not be the case if the balance sheet is compiled with an electronic spreadsheet that has not been properly configured. If so, the person compiling the budget may elect to manually insert the difference between the two totals in a 「plug」 account, such as other assets or other liabilities. If the amount of this plug is material, it can call into question the reliability of the information in the budgeted balance sheet, and so may trigger an examination of the assumptions and formulas used to construct the balance sheet.

Figure 7.11 is an example of budgeted balance sheet.

Very Large Corporation
Budgeted Balance Sheet
As of Year December 31, 20XX

Current Assets			
Cash	(1)	$1,500,000	
Accounts receivable	(2)	4,200,000	
Raw materials inventory	(3)	3,500,000	
Finished goods inventory	(4)	6,800,000	
Total Current Assets			$16,000,000
Fixed Assets			
Office equipment	(5)	500,000	
Machinery	(6)	9,200,000	
Accumulated depreciation	(7)	-2,700,000	
Net Fixed Assets			7,000,000
Total Assets			$23,000,000
Current Liabilities			
Accounts payable	(8)	$2,100,000	
Notes payable	(9)	5,900,000	
Total Current Liabilities			8,000,000
Shareholders' Equity	(10)		15,000,000
Total Liabilities & Equity			$23,000,000

Figure 7.11　An Example of Budgeted Balance Sheet

Notes to the budgeted balance sheet example:
a) Reference from the ending cash balance noted on the cash budget.
b) 10% of third quarter sales plus 40% of fourth quarter sales.
c) Reference from the direct materials budget.
d) Reference from the ending finished goods inventory budget.
e) Reference from the capital spending budget.

f) Reference from the capital spending budget.
g) Reference from the capital spending budget.
h) Reference of all expenses in the selling and administrative expense budget, the direct materials budget, and the manufacturing overhead budget, assuming 30-day payment terms.
i) Reference from the financing budget.
j) Reference from the last actual balance sheet, plus projected budgeted net earnings.

7.5 Using Information Technology for Sensitivity Analysis, Variance Analysis and Rolling Up Unit Budgets

Figure 7.4 through figure 7.11 show that the manager must prepare many calculations to develop the master budget for just one product. No wonder managers embrace information technology to help prepare budgets. Let us see how advances in information technology make it more cost-effective for managers to:
a) Conduct sensitivity analysis on their own unit's budget.
b) Roll up individual unit budgets to create the company-wide budget.

7.5.1 Sensitivity Analysis

敏感性分析: What if 分析, 即當不確定因素變動時, 對指標的影響程度和敏感性。

The master budget models the company's planned activities. Top managers pay special attention to ensure that the results of the budgeted income statement, the cash budget, and the budgeted balance sheet support key strategies.

But actual results often differ from plans, so manager wants to know how budgeted income and cash balances would change if key assumptions were changed. In previous chapter, we defined sensitivity analysis as a what-if technique that asks what a result will be if a predicted amount is not achieved or if an underlying assumption changes. What is the stock market crashed? How will this affect company's sales? Will it have to postpone the planned expansion in Asia and Europe?

What will be a company's cash balance on Dec 31st if the uncollectible sales increase from 1% to 2%? Will the company have to borrow more cash? If a company increases sales revenue, can it generate net income?

Most companies use computer spreadsheet programs to prepare master budget schedules and statements. One of the earliest spreadsheet programs was developed by graduate business students who realized that computers could take the drudgery out of hand-computed master budget sensitivity analyses. Today, managers answer what-if questions simply by changing a number. At the press of a key, the computer screen flashed a revised budget that includes all the effects of the change.

Technology makes it cost-effective to perform more comprehensive sensitivity analyses. Armed with a better understanding of how changes in sales and costs are likely to affect the company's bottom line, today's managers can react quickly if key assumption underlying the master budget (such as sales price or quantity) turn out to be wrong.

7.5.2 Variance Analysis

差異分析:將實際結果與預算比較,計算其差異,分析差異產生的原因。

Budgets are prepared for specific periods so that managers can compare actual results for the period with the planned results for that period. Variance analysis has many forms and can result in complex measures, but, as shown in figure 7.12, its basis is very simple—an actual cost or actual revenue amount is compared with a target cost or target revenue amount to identify the difference, which is called a variance. For example, a manager might compute the cost of labor that went into making an aircraft and compare that cost with the planned cost of labor for making that aircraft. A variance represents a departure from what was budgeted or planned. What caused the variance and the size of that variance will trigger an investigation to determine its cause and what should be done to correct that variance. Budgeted, or planned, costs can come from three sources:

a) Standards established by industrial engineers, such as cost of steel that should go into an automobile door based on the door's specifications.

b) Previous period's performance, such as the average cost of steel per door that was made in the last budget period.

c) A performance level achieved by a competitor—usually called a benchmark and based on best-in-class results—such as the cost of steel per comparable door achieved by a competitor who is viewed as the most efficient.

```
┌─────────────────┐         ┌─────────────────┐
│  Actual cost    │ ⇐====⇒ │  Budgeted cost  │
│  $12,350,000    │         │  $12,530,000    │
└─────────────────┘         └─────────────────┘
                   ⇓
            ┌──────────────────┐
            │ Variance$180,000 │
            └──────────────────┘
```

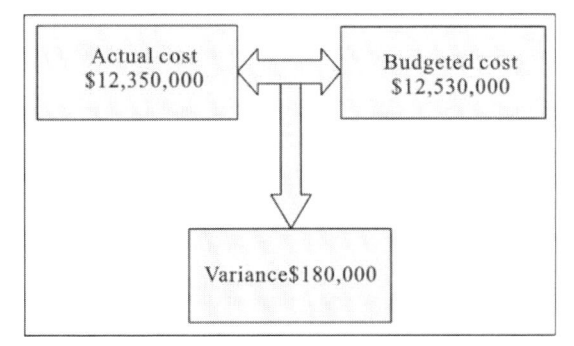

Figure 7.12 Variance Analysis

The financial numbers used in variance analysis for flexible resources are the product of a price and a quantity component:

Planned, or budgeted, amount
= Standard price per unit × Budgeted quantity
While,
Actual amount = Actual price per unit × Actual quantity

Variance analysis explains the difference between planned costs and actual costs by evaluating differences between standard prices and actual prices and budgeted quantities and actual quantities. Managers focus separately on prices and quantities because in most organizations one department or division is responsible for the acquisition (thereby determining the actual price) of a resource and a different department uses (thereby determining the quantity of) the resource.

A variance is a signal that is part of a control system for monitoring results, and thus variances provide a signal that operations did not go as planned. Supervisory personnel use variances as overall checks on how well the people who are managing day-to-day operations are doing what they should be doing. When compared to the performance of other organizations engaged in comparable tasks, variances show the effectiveness of the control systems that operations people are using.

7.5.3 Rolling Up Individual Unit Budgets into the Company-wide Budget

Many companies' headquarters must roll up the budget data from all product divisions to prepare the company-wide master budget. This roll up can be difficult for companies

總部需將所有部門的預算數據匯總，並據以編制企業的總預算。

whose units use different spreadsheets to prepare the budgets.

Lots of companies decided to use budget-management software to solve this problem. Often designed as a component of the company's enterprise resource planning (ERP) system (or date warehouse), this software helps managers develop and analyze budgets.

　　企業資源計劃(ERP)是一企業訊息管理的系統軟件。

Across the globe, managers sit at their desks, log into the company's budget system, and enter their budget amounts. The software allows them to conduct sensitivity analyses on their own unit's date. When the manager is satisfied with the budget, it can be entered in the company-wide budget with the click of a mouse. The specific unit's budget automatically rolls up with budgets from all other units around the world.

Whether at headquarters or on the road, top executives can log into the budget system and conduct their own sensitivity analyses on individual units' budgets or on the company-wide budget. Managers can spend less time complying and summarizing data have more time analyzing it to ensure that the budget leads the company to achieving its key strategic goals.

QUESTIONS:

1. The correlation coefficient (r) for measuring the connection between two variables (x and y) has been calculated as 0.6.
How much of the variation in the dependent variable (y) is explained by the variation in the independent variable (x)?

A. 36%　　　　　　　　　　B. 40%
C. 60%　　　　　　　　　　D. 64%

2. A company uses regression analysis to establish its selling overhead costs for budgeting purposes. The data used for the analysis is as follows:

Month	Number of salesmen	Sales overhead costs $000
1	3	35.1
2	6	46.4
3	4	27.0
4	3	33.5
5	5	41.0
	21	183.0

The gradient of the regression line is 4.20. Using regression analysis, what would be the budgeted sales overhead costs for the month, in $000, if there are 5 salesmen employed?
A. 28.87
B. 39.96
C. 41.00
D. 56.76

3. The main purposes of budgeting are:
(i) to give authority to spend.
(ii) to control expenditure.
(iii) to aid decision making.
A. (i) only
B. (i) and (ii) only
C. (ii) only
D. (i), (ii) and (iii)

4. A business is preparing its production budget for the year ahead for product A998. It is estimated that 100,000 units of A998 can be sold in the year and the opening inventory is currently 14,000 units. The inventory level is to be reduced by 40% by the end of the year.
The number of units of A998 needed to be produced is:
A. 86,000
B. 94,400
C. 100,000
D. 108,400

5. A process has a normal loss of 10% and budgeted output is 4,500 litres for the period. Opening inventory of raw material is 600 litres and is expected to increase by 20% by the end of the period.
The material usage budget is:
A. 4,500 litres
B. 5,000 litres
C. 5,133 litres
D. 5,120 litres

6. A company makes three products, X, Y and Z. The following information is available:

	X	Y	Z
Budgeted production (units)	200	400	300
Machine hours per unit	5	6	2
Variable overheads	$2.30 per machine hour		
Fixed overheads	$0.75 per machine hour		

The overhead budget is:
A. $12,200
B. $12,000

C. $11,590
D. $10,980

7. The material usage budget is calculated by taking the production budget and multiplying by the standard material quantity per unit.
This statement is:
A. True
B. False

8. A job requires 2,400 actual labour hours for completion but it is anticipated that idle time will be 20% of the total time required. If the wage rate is $10 per hour, what is the budgeted labour cost for the job, including the cost of the idle time?
A. $19,200
B. $24,000
C. $28,800
D. $30,000

9. A company has a budget for two products A and B as follows:

	Product A	Product B
Sales (units)	2,000	4,500
Production (units)	1,750	5,000
Labour:		
Skilled at $10/hour	2 hours/unit	2 hours/unit
Unskilled at $7/hour	3 hours/unit	4 hours/unit

What is the budgeted cost for unskilled labour for the period?
A. $105,000
B. $135,000
C. $176,750
D. $252,500

10. A company makes two products, X and Y, which are sold in the ratio 1 : 2. The selling prices are $50 and $100 respectively. The company wants to earn $100,000 over the next period. The sales budget should be:

	X (units)	Y (units)
A.	1,334	667
B.	800	400
C.	667	1,334
D.	400	800

Chapter 8 Standard Costing and Variance Analysis

Learning Objectives

After the study of this chapter, you should be able to:
a) Understand the purpose of standard costing and how standard costing operates.
b) Explain how standard costs are set.
c) Calculate material, labor, overhead and sales profit variance and reconcile actual profit with budgeted profit.
d) Distinguish between standard variable costing and standard absorption costing.

8.1 Standard Costing Overview

The actual results achieved by an organization during a reporting period will more than likely, be differ from the expected results which standard are set by budget that we learned in the previous chapter.

標準成本：預先設定的目標成本。

A standard cost is a predetermined unit cost and can be a unit cost of production or unit cost of a service. Standard costing which is used with a system of budgeting is a control system designed to enable the deviations from budget to be analyzed in detail, thus enabling costs to be controlled more effectively. In this chapter we will examine how standard costing system operates and how the variances are calculated.

標準成本法與預算系統一起使用，是管理控制系統的一部分。

Standard costing is most suited to an organization whose activities consist of a series of common or repetitive operations and the input required to produce each unit of output can be specified. It is therefore relevant in manufacturing companies, since the process involved are often of a repetitive nature. The figure 8.1 below provides an overview of a standard costing system:

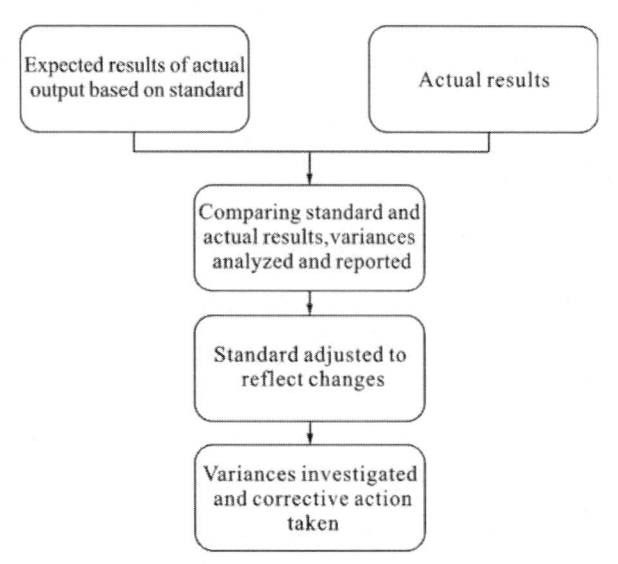

Figure 8.1 Standard Costing System

Budgetary control is the process of managers being responsible for their budget and with this responsibility there is the problem of trying to find the differences between the budget and the actual figures. This is exactly where standard costing helps these managers discover what the difference is and in which area the difference occurs.

After cost standard for each unit produced set, manager will compare actual results with expected results, variance may occur between individual items, such as the cost of material and the volume of sales, wise manager will use the comparison information to assist in making decision whether to carry out the investigations to identify the reasons for the variance, and then take appropriate action or review standards.

8.2 Establishing Cost Standard

Standard costs can be a unit cost of production or unit cost of a service. They can be set based on either historical experience or engineering studies.

設定標準成本可參考歷史數據或數學模型。

The setting of the standards is a very difficult task as the standard setter has to predict what might happen in the next year. For example, setting the raw material cost standard requires information from many sources, such as purchase department, production department, accounting department and market forecast by work study.

Although average historical records are widely used to set standard in practice, it should be used with caution because the latter might include past inefficiencies. Labor cost is an example that is likely to contain inefficient work in the past, and should be renewed to eliminate the inefficiencies. Most of the standard are only set once a year and are used throughout the coming year even if there are major changes in the underlying cost base due to unforeseen circumstances, e.g. an increase in inflation rate.

In some situations, the management has little influence on sales price which is mainly driven by market force. Obviously it makes sense that they set up the desirable profit margin first, then the target cost is determined by deducting desirable profits from expected sales.

Example 8.1

Suppose one product A can be sold at $10 in the market. If the company wants 12% profit, the target cost per unit should be:

Target cost per unit = 10×(1−12%) = 8.8

8.2.1 Type of Standard

標準成本有三種類型：基本標準、理想標準、可達標準。

Managements generally set standards for costs based on three levels of attainment.

(1) Basic Standard

Basic cost standards represent constant standards that are left unchanged over long periods. They are used to show the changes in efficiency or performance over a long period of years. When changes occur in methods of production, price level or other relevant factors, basic standard are not very useful.

(2) Ideal Standard

Represent perfect performance, no wastage, no spoilage, no inefficiencies, no idle time, no breakdowns, ideal standard costs are the minimum costs that are possible under the most efficient operating conditions. However, ideal standards are unlikely to be used in practice because they may have an adverse impact on employee motivation.

(3) Currently Attainable Standard

These standards represent those costs that should be incurred under efficient operating condition. They are standards that are difficult, but not impossible to achieve because some allowance are made for wastage and inefficiencies. The fact that these standards represent a target that can be achieved under efficient conditions, but which is also viewed as being neither too easy to achieve nor impossible to achieve, provides the best norm to which actual costs should be compared.

8.2.2 Standard Cost Card

After establishing standard costs for the specific operation / product, it is presented in the standard cost card.

Example 8.2

Beta Ltd. produces a single product, which is known as sigma. The entire product is sold as soon as it is produced (in other words, no opening nor closing inventory). The product requires a single operation, and the standard cost for this operation is presented in table 8.1 as below:

Table 8.1　　　　　**Standard Cost Card for Product Sigma**

Standard cost card for product sigma	$
Direct material	
2 kg of A at $10 per kg	20
1 kg of B at $15 per kg	15
Direct labor (3 hours at $9 per hour)	27
Variable overhead (3 hours at $2 per direct labor hour)	6
Standard variable cost of production	68
Fixed production overhead	12
Standard full cost of production	80
Standard gross profit	8
Standard selling price	88

Note: Budgeted fixed overheads are $1,440,000 and are assumed to be incurred evenly throughout the year, budgeted production is annual 120,000units.

8.2.3 How Standard Cost Cards Are Used

標準成本的應用:存貨的計量和成本控制。

Firstly, standard cost is used to value inventories and cost production for cost accounting purpose, in which case it is easy and means that all issues to production are at the standard so production jobs can be compared. The use of standard price for material is helpful in most situations but managers should keep in minds that reconciling the actual cost of materials with the standard cost of materials is often necessary.

定期將實際結果與標準比較，兩者之間的差額稱為「差異」，分為有利差異(F)與不利差異(A)。

The other purpose, is that standard costs (planned, budgeted, expected, estimated) act as a control device, it should be periodically compared to actual results and the differences are calculated. These differences are referred to as variances. If actual results are better than expected results, favorable variance (F) incurs, otherwise, adverse variance (A) emerges.

Assuming Beta Ltd plans to manufacture 10,000 units of sigma in the month of March. Budget based on the above standard costs and an output of 10,000 units, please look at table 8.2 and table 8.3 as following:

Table 8.2   **Budgeted results for March ($)**

Sales (10,000 units of sigma at $88 per unit)		880,000
Direct material		
A: 20,000 kg at $10 per kg	200,000	
B: 10,000 kg at $15 per kg	150,000	350,000
Direct labor (30,000 hr at $9 per hr)		270,000
Variable production overheads (30,000 hr at $2 per direct labor hr)		60,000
Total variable production overheads		(680,000)
Fixed overheads		(120,000)
Budgeted profit		80,000

Note: Budgeted fixed overheads are $1,440,000 and are assumed to be incurred evenly throughout the year.

Table 8.3   **Actual results for March ($)**

Actual sales (9,000 units at $90)		810,000
Actual direct materials:		
A: 17,800 kg at $11 per kg	195,800	
B: 10,100 kg at $14 per kg	141,400	337,200
Actual direct labor (28,500 hours at $9.6 per hr)		273,600
Actual variable production overheads		52,000
Total variable production overheads		(662,800)
Actual fixed overheads		(116,000)
Actual profit		31,200

Next we will learn how the variances are calculated, note that the following variances calculation are illustrated from the information contained in example Beta Ltd.

8.3 Basic Variance Analysis

標準成本法下, 總的成本差異分為價格差異和數量差異。

In standard costing system, total variance is broken down into price and usage variances. Price (rate) variance is the difference between the actual and standard unit price of an input multiplied by the number of inputs used. Usage variance is the difference between the actual and standard quantity of inputs multiplied by the standard unit price of the input. It is easy to show that the total variance is the sum of price and usage variances:

$$\text{Total variance} = \text{Price variance} + \text{Usage variance}$$

This computation must be done for every class of direct material and every class of direct labor. The treatment of overhead is discussed later.

8.3.1 Direct Material Cost Variance

The costs of materials which are used in a manufactured product are determined by two basic factors: the unit price paid for the materials, and the quantity of materials used in production. Therefore, we can calculate the total material variance by compute two constituted variance.

Direct material price variance is the difference between actual and standard unit price multiplied by the actual quantity input.

原材料成本差異: 由價差與量差構成。

Direct material usage is the difference between actual and standard quantity of inputs multiplied by standard unit price.

Direct material total variance
= Direct material price variance + Direct material usage variance

Now refer to material A in example Beta Ltd. Please look at table 8.4, table 8.5, table 8.6 and table 8.7 as following.

Table 8.4 Direct Material A Total Variance

	$	
Producing 9,000 units sigma should have cost	180,000	
Did cost	195,800	
Total material A variance	15,800	A

Table 8.5   Direct Material A Price Variance

	$	
17,800kg should have cost ($10/kg)	178,000	
Did cost	195,800	
Material A price variance	17,800	A

Table 8.6   Direct Material A Usage Variance

9,000 actual units should have used (kg)	18,000	
But did use (kg)	17,800	
Difference in usage (kg)	200	
×Standard unit price ($/kg)	× $10	
Material A usage variance ($)	$2,000	F

Table 8.7   Summary

	$	
Price variance	17,800	A
Usage variance	2,000	F
Total variance	15,800	A

Practice

Based on the data given in Beta Ltd, calculate the variance of material B, including price and usage variance.

8.3.2   Direct Labor Cost Variance

人工成本差異：工資率差異與效率差異。

The cost of labor is determined by the price (rate) paid for labor and the quantity of labor (efficiency) used. Thus,

Total direct labor variance = Labor rate variance + Labor efficiency variance

The labor rate variance, which is similar to the material price variance, is the difference between the standard cost and the actual cost for the actual number of hours worked.

The labor efficiency variance, which is similar to the material usage variance, is the difference between the hours that should be used for the actual output, and the actual number of hours worked, in terms of standard rate per hour.

Now referring to labor cost in example Beta Ltd. Please look at table 8.8, table 8.9, table 8.10 and table 8.11 as following:

Table 8.8　　　　Direct Labor Total Variance

	$	
Producing 9,000 units sigma should have cost	243,000	
Did cost	273,600	
Total labor variance	30,600	A

Table 8.9　　　　Direct Labor Rate Variance

	$	
28,500 hours should have cost ($9)	256,500	
Did cost	273,600	
Labor rate variance	17,100	A

Table 8.10　　　　Direct Labor Efficiency Variance

9,000 actual units should have used (hour)	27,000	
But did use (hour)	28,500	
Difference in usage (hour)	1,500	
×Standard labor rate per hour ($/hour)	× $9	
Labor efficiency variance ($9)	$13,500	A

Table 8.11　　　　Summary

	$	
Labor rate variance	17,100	
Labor efficiency variance	13,500	
Total variance	30,600	A

8.3.3　Variable Overhead Variance

變動制造費用差異可分為耗費差異和效率差異。

Here comes to overhead variance analysis. First, we will divide overhead into two categories: variable and fixed. Next, we will look at variances for each category.

The total variable overhead variance is calculated in the same way as the total direct labor variances. It is divided into two components: the variable overhead expenditure variance and variable overhead efficiency variance.

Based on actual units of production, total variance is the difference between the standard variable overhead charged to product and the actual variable overhead incurred.

Where variable overhead vary with direct labor hour of input, the total variable overhead variance 2,000(F) may be due to one or both of the following factors:

a) a price variance arising from actual expenditure being different from budgeted expenditure

b) a quantity variance arising from actual direct labor of input being different from the hours of input, which should have been used.

Please look at table 8.12, table 8.13, table 8.14 and table 8.15 as following:

Table 8.12   Total Variable Overhead Variance

	$	
Producing 9,000 units sigma should have cost	54,000	
Did cost	52,000	
Total variable overhead variance	2,000	F

Table 8.13   Variable Overhead Expenditure Variance

	$	
28,500 hours should have cost ($2)	57,000	
Did cost	52,000	
Labor rate variance	5,000	F

Table 8.14   Variable Overhead Efficiency Variance

9,000 actual units should have taken (hour)	27,000	
But did use (hour)	28,500	
Difference in hours (hour)	1,500	
×Standard variable overhead rate per hour ($/hour)	×$2	
Labor efficiency variance ($)	$3,000	A

Table 8.15   Summary

	$	
Expenditure variance	5,000	F
Efficiency variance	3,000	A
Total variance	2,000	F

8.3.4 Fixed Overhead Variance

We will again use the Beta Ltd. example to illustrate the computation of the fixed overhead variances, which is a very different method to what we have learned in calculating variable cost variances.

There are three elements to consider when we come to analysis fixed overhead variance.

固定製造費用總差異是實際產量下的成本差異。

(i) Actual fixed overhead with actual outputs.
(ii) Budgeted fixed overhead with budgeted outputs.
(iii) Budgeted fixed overhead with actual outputs.

Fixed overhead variances are the costs variance with actual units, that is

Fixed overhead variance = (i) - (ii)

固定製造費用總差異也是在完全成本法中,過度吸收/吸引不足的部份。

In absorption costing system, fixed overhead variances are an attempt to explain the under-absorption or over-absorption of fixed overheads in production costs. We look at under/over absorption of fixed overhead in previous chapter and know:

Fixed overhead variance = under-absorption or over-absorption of fixed overhead
= (i) - (iii)

In Beta Ltd. example, budgeted monthly fixed overhead is $120,000 ($1,440,000 annually), and budgeted activity is 10,000 units for March.
Since the actual activity of March is 9,000 units,
Fixed overhead that has been charged to products = $12 × 9,000 = $108,000
Comparing to actual fixed overhead of $116,000, there is $8,000 under absorbed fixed overhead. The under / over absorption of fixed overhead represents the total fixed overhead variance for the period.

Total variance = (i) - (iii)
= Actual fixed overhead - Budgeted fixed overhead with actual output

Total variance (under /over absorption) may be due to two factors, expenditure variance and volume variance:

耗費差異:預算費用與實際費用的差異。

a) Fixed overhead expenditure variance of $40,000 arising from actual expenditure ($116,000) being different from budgeted expenditure ($120,000).

Expenditure variance = (i) - (ii)
= Actual fixed overhead - Budgeted fixed overhead with budgeted outputs

數量差異:實際產量與預算產量的差異。

b) Fixed overhead volume variance arising from actual production differing from budgeted production.

Volume variance = (ii) − (iii)
(Budgeted outputs − Actual outputs) × Standard overhead absorption rate

Thinking: Why the actual volume differ from budgeted volume?

效率差異:員工工作效率與預算不一致。

a) One possible reason is that the labor force worked at a different level of efficiency from that anticipated in budget. In Beta scenario, the labor force had worked for 28,500 hours to produce 9,000 units which have expected to required 27,000 hours. Thus the inefficiency of labor is one of the reasons why the actual production was less than the budgeted production. It is volume efficiency variance that can be computed by:

Volume efficiency variance = (Standard hours of output − Actual hours of input) × Standard overhead absorption rate per hour

能力差異:實際利用生產能力與預算不一致。

b) The second reason is that the actual hours worked are only 28,500 hours but the budget planed 30,000 hours for use, perhaps was caused by shortage of labors. The difference which is measured by volume capacity variance, of 1,500 hours reflects the fact that the company has failed to utilize the planned capacity.

Volume capacity variance = (Budgeted hour of input − Actual hour of input) × Standard overhead absorption rate per hour

Please look at table 8.16 ~ Table 8.21 as following:

Table 8.16   Fixed Overhead Total Variance (Under / over absorption)

	$	
Fixed overhead incurred	116,000	
Fixed overhead absorbed (9,000units ×12)	108,000	
Fixed overhead total variance (under/over absorbed overhead)	8,000	A

Table 8.17   Fixed Overhead Expenditure Variance

	$	
Fixed overhead incurred	116,000	
Budgeted fixed overhead	120,000	
Fixed overhead expenditure variance	4,000	F

Table 8.18　　　　　Fixed Overhead Volume Variance

		$	
Actual output	9,000 units		
Budgeted output	10,000 units		
Difference	1,000 units		
×Standard overhead absorption rate	× $12 per unit		
Volume variance		12,000	A

Table 8.19　　　　　Fixed Overhead Efficiency Variance

		$	
9,000 units should take	27,000 hours		
Did take	28,500 hours		
Difference	1,500 hours	A	
×Overhead absorption rate per hour	× $4 per hour		
Volume efficiency variance		6,000	A

Table 8.20　　　　　Fixed Overhead Capacity Variance

		$	
Budgeted hours of work	30,000 hours		
Actual hours of work	28,500 hours		
Difference	1,500 hours	A	
×Overhead absorption rate per hour	× $4 per hour		
Volume efficiency variance		6,000	A

Table 8.21　　　　　Summary of Fixed Overhead Variances

	$	
Expenditure variance	4,000	F
Efficiency variance	6,000	A
Capacity variance	6,000	A
Total variance (under-absorbed)	8,000	A

8.3.5　Summary of Cost Variance Analysis

(1) Costs Variance Analysis

Fixed overheads are assumed to remain unchanged in the short term despite the changes in activity level, but they may change in response to other factors.

Please look at figure 8.2 and figure 8.3.

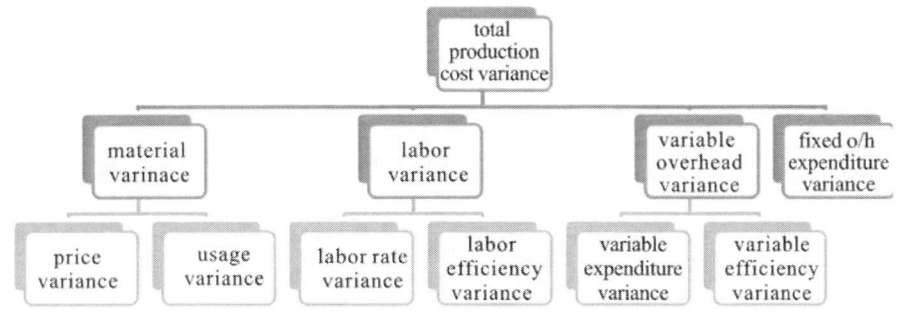

Figure 8.2 Costs Variance Analysis(1)

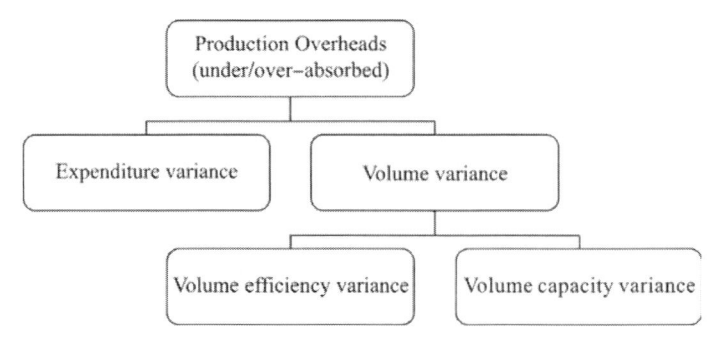

Figure 8.3 Costs Variance Analysis(2)

(2) Cost Control-Possible Causes for Cost Variances

In practice, variances occur all the time due to many reasons, even by inaccurate recording of actual costs. Managers should be able to identify whether the variance is caused by operating inefficiency or inappropriate standard and react accordingly.

There are many possible reasons for cost variances arising, as you will see from the following list of possible causes(Table 8.22):

Table 8.22   Possible Reasons for Cost Variances Arising

Variance	Favorable	Adverse
Material price	Unexpected discounted received; More care taken in purchasing; Change in material standard	Price increase, careless purchasing, change in material standard
Material usage	Higher-quality material used, more effective use made of material, minimizing scrap	Defective material, excessive waste
Labor rate	Use of workers at a rate of pay lower than standard	Unexpected overtime, labor rate increase, use of skilled worker

Table8. 22 (continued)

Variance	Favorable	Adverse
Labor efficiency	Higher output due to work motivation, better quality of equipment or material, or better method	Lower output because of lack of training, sub-standard material used, or deliberate restriction
Overhead expenditure	Saving in costs incurred; More economical use of service	Excess use of services
Overhead volume efficiency	Labor force working more efficiently	Labor force working less efficiently
Overhead volume capacity	Labor force working overtime; Level of activity greater than budgeted	Machine breakdown; Labor / material shortage; Poor production scheduling; Reduction in sales demand

(3) Significance of Cost Variance

Due to the complex business environment, actual performance rarely exactly meets the standards and random variations around the standard are expected. Manager should not look at all variances because it is time consuming and expensive, instead, the management time is spent only in considering the larger variances and there can be a concentrated effort placed on finding the reasons for these variances.

How do managers determine whether variances are worth further investigated? There are a number of factors which can be taken into account.

重要性。

a) Materiality. Many firms do not investigate variances unless they are large enough to be of concern, more specifically, falling outside of an acceptance range. Some firms set control limit that expressed both as a percentage of the standard and as an absolute dollar amount, say 5% above or below the expected costs, whereas others might use judgement rather than any identification of limits.

可控性。

b) Controllability. Some types of variances may not be controllable even once their cause is revealed.

For example, labor rate, which are largely determined by external factors such as labor market, is probably the one that is least subject to control by management. When labor rate variance occurs, in most cases the variances is due to the standards not being kept in line with changes in actual labor rates.

Uncontrollable variances call for a change in the standard,

各因素之間的相關性。

not an investigation into the past.

c) Interdependence between Variance. Individual variances should not be looked at in isolation. One variance might be inter-related with another, and more likely be caused by the other. Here are some examples, please look at table 8.23 as following:

Table 8.23　　The Example of Interdependence Between Variance

Interrelated variances	Explanation
Material price and usage	Imagine that the purchasing department purchases cheap but sub-quality material with favorable material price variance, this may lead to inferior product quality or more waste, thus arises excessive material usage and / or labor inefficiency. If so, the adverse material usage variance and adverse labor efficiency variance should be charged to the purchasing department instead of production department
Labor rate and efficiency	Higher rate paid for experience and skill likely leads to adverse labor rate variance but favorable efficiency variance
Selling price and sales volumes	A reduction in the selling price might stimulate bigger sales demand, so that an adverse selling price variance might be counterbalanced by a favorable sales volume variance

8.4　Sales Variance

Other than costs, sales are another important factor for achieving planed profit and sales variance can be used to analyze the performance of business. The most significant feature of sales variance calculations is that they are calculated in terms of profit or contribution margins rather than sales value.

銷售額差異:單價差異和銷量差異。

Sales are determined by unit price and sales volume. Thus two variances must be calculated as indicated below:

Selling price variance = Difference between planned and actual unit price × Actual unit sold
It measures how much the changes in selling price affect profit.

Selling volume profit variance = Difference between actual unit sold and the planned quantity × Planned profit per unit
It measures the effect of changes in sales volume on actual profit.

Again we use Beta Ltd. to illustrate how the sales price variance and sales volume are calculate, please look at table 8.24, table 8.25 and figure 8.4 as following:

Table 8.24   **Selling Price Variance**

	$	
Sales revenue from selling 9,000 units sigma should be (standard selling price of $88 / unit)	792,000	
Actually was (actually was $90 / unit)	810,000	
Selling price variance	18,000	F

Table 8.25   **Sales Volume Profit Variance**

		$	
Budgeted sales volume	10,000 units		
Actual sales volume	9,000 units		
Difference in sales volume	1,000 units	A	
×budgeted profit per unit	× $8 per unit		
Volume efficiency variance		8,000	A

Budgeted profit per unit = Unit selling price − Full cost per unit
= 88 − (68 + 12) = 8

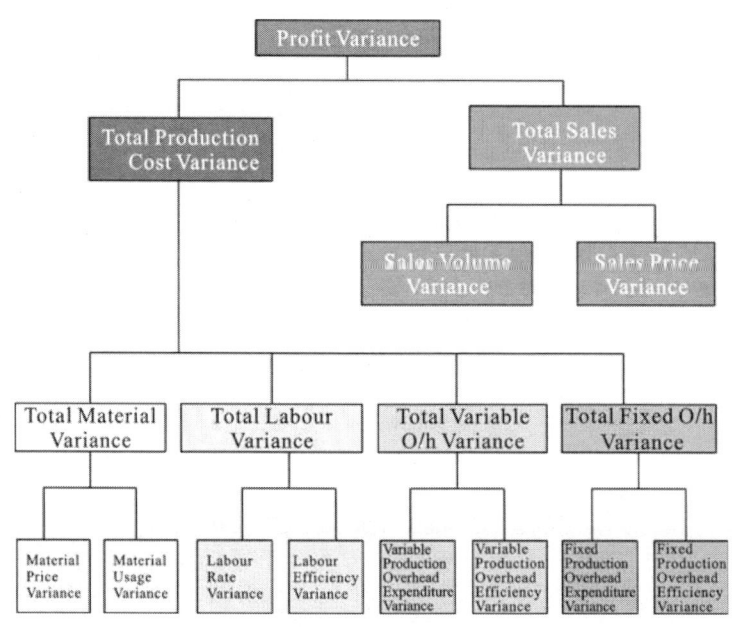

Figure 8.4   Summary

8.5 Reconciliation of Budget and Actual Profit Under Standard Absorption Costing

完全成本法下,預算利潤與實際利潤差異分析。

根據成本費用收入差異,編制業務報表。

Standard costing may be used either in absorption cost or marginal costing. Management will be interested in the reason for the actual profit being different from the budgeted profit. This information is shown in the reconciliation statement (also called operating statement) in respect of example Beta Ltd. Please look at table 8.26 as following:

Table 8.26 Beta – Reconciliation Statement for March

Beta – Operating statement for March	$		$		$	
Budgeted profit for March					80,000	
Sales variances:						
Sales price variance	18,000	F				
Sales volume variance	8,000	A			10,000	F
Actual sales minus standard cost of sales					90,000	
Cost variance						
Material price – A	17,800	A				
– B	10,100	F				
Material usage – A	2,000	F				
– B	16,500	A	22,200	A		
Labor rate	17,100	A				
Labor efficiency	13,500	A	30,600	A		
Variable overhead expenditure	5,000	F				
Variable overhead efficiency	3,000	A	2,000	F		
Fixed overhead expenditure	4,000	F				
Fixed overhead volume efficiency	6,000	A				
Fixed overhead volume capacity	6,000	A	8,000	A	58,800	A
Actual profit for March					31,200	

8.6 Reconciliation of Budget and Actual Profit or Contribution Under Standard Marginal Costing

邊際成本法下,預算利潤與實際利潤的差異分析。

If a firm uses standard marginal costing instead of standard absorption costing, there will be two differences in the way the variances are calculated.

a) There is no fixed overhead volume variance. In marginal costing, fixed costs are not allocated to product and so there are no fixed cost variances to explain any under/over-absorption of overhead. Therefore, there is no fixed overhead volume variance.

b) The sales volume variance is valued at standard contribution margin, not standard profit margin.

In Example Beta Ltd.

sales volume variance = Difference between actual and budgeted unit sold × Standard contribution margin
= 1,000 × 20
= 20,000 (A)

根據成本費用收入差異,編制業務報表。

Table 8.27　　　　Beta-Reconciliation Statement for March

Beta-Operating statement for March	$	$	$	
Budgeted contribution			200,000	
Sales variances:				
Sales price variance	18,000	F		
Sales volume variance	20,000	A	2,000	A
Actual sales minus the standard variable cost of sales			198,000	
Variable cost variance				
Material price - A	17,800	A		
- B	10,100	F		
Material usage - A	2,000	F		
B	16,500	A	22,200	A
Labor rate	17,100	A		
Labor efficiency	13,500	A	30,600	A
Variable overhead expenditure	5,000	F		
Variable overhead efficiency	3,000	A	2,000	F
			50,800	A
Actual contribution			147,200	
Budgeted fixed overhead		120,000		
Fixed overhead expenditure variance		4,000	F	
Actual fixed overhead			116,000	
Actual profit for March			31,200	

Practice

Q plc uses a standard costing system. The standard unit cost card is as follows(table 8.28):

Table 8.28   The standard Cost Card

Selling price per unit	$80
Direct material (4 kg at $2/kg) per unit	$8
Direct labor (4 hours at $4/hour) per unit	$16
Variable overhead (4 hours at $3/hour) per unit	$12
Fixed production overhead (4 hours at $5/hours)	$20
Administration costs	$10,000
Monthly output	500 units

Fixed overheads are absorbed on the basis of labor hours. The actual result for April is as follows(table 8.29):

Table 8.29   Actual Result

Selling price per unit	$82
Direct material (2,100kg at $1.8/kg)	$3,780
Direct labor (1,950 hours at $5/hour)	$9,750
Variable overhead (4 hours at $3/hour)	$6,380
Fixed production overhead incurred	$9,800
Administration costs	$9,800
Monthly output for April	480 units

Assuming there is no opening and closing inventory for April.

Required

Prepare the operating statement for April (using absorption costing and marginal costing respectively) to show the budgeted profit, variances for material, labor, overhead and sales, and actual profit.

QUESTIONS

1. Which of the following statement is more likely to be true?
A. Basic standards may discourage employees in the way they frequently lead to adverse variances
B. Current standard represent the level of productivity which management will wish to plan for
C. Once the standard have been set, it must be remained unchanged throughout the coming budget period despite the unexpected high inflation

D. Ideal standard may provide an incentive to greater efficiency even though the standard cannot be achieved

2. A company uses standard absorption costing. The following data relate to last month:

	Budget	Actual
Sales and production (units)	1,000	900
Standard Actual		
	$	$
Selling price per unit	50	52
Total production cost per unit	39	40

What was the adverse sales volume profit variance last month?
A. $1,000
B. $1,100
C. $1,200
D. $1,300

3. Under absorption costing principles a favourable sales volume variance is calculated as the difference in sales volumes multiplied by:
A. Standard contribution per unit
B. Standard cost per unit
C. Standard profit per unit
D. Standard selling price per unit

4. FGH has the following budgeted and actual data:
Budgeted fixed overhead cost $120,000
Budgeted production (units)   20,000
Actual fixed overhead cost $115,000
Actual production (units)   21,000
The fixed overhead volume variance:
A. is $4,500 adverse
B. is $5,500 favourable
C. is $6,000 favourable
D. is $10,500 favourable

5. A company operates a standard marginal costing system. Last month its actual fixed overhead expenditure was 10% above budget resulting in a fixed overhead expenditure variance of $36,000.
What was the actual expenditure on fixed overheads last month?
A. $324,000
B. $360,000
C. $396,000
D. $400,000

6. A company uses standard marginal costing. Last month the standard contribution on actual sales was $10,000 and the following variances arose:
Total variable costs variance $2,000 adverse
Sales Price variance $500 favourable
Sales volume contribution variance $1,000 adverse
What was the actual contribution for last month?
A. $7,000
B. $7,500
C. $8,000
D. $8,500

7. A company uses standard marginal costing. Last month, when all sales were at the standard selling price, the standard contribution from actual sales was $50,000 and the following variances arose:
Total variable cost variance $3,500 adverse
Total fixed costs variance $1,000 favourable
Sales volume contribution variance $2,000 favourable
What was the actual contribution for last month?
A. $46,500
B. $47,500
C. $48,500
D. $49,500

8. The following information relates to labour costs for the past month:

Budget	Labour rate	$10 per hour
	Production time	15,000 hours
	Time per unit	3 hours
	Production units	5,000 units
Actual	Wages paid	$176,000
	Production	5,500 units
	Total hours worked	14,000 hours

There was no idle time.
What were the labour rate and efficiency variances?

	Rate variance	Efficiency variance
A.	$26,000 adverse	$25,000 favourable
B.	$26,000 adverse	$10,000 favourable
C.	$36,000 adverse	$2,500 favourable
D.	$36,000 adverse	$25,000 favourable

9. A company budgeted to make 30,000 units of a product P. Each unit was expected to take 4 hours to make and budgeted fixed overhead expenditure was $840,000. Actual production of product P in the period was 32,000 units, which took 123,000 hours to make. Actual fixed overhead expenditure was $885,600.

What was the fixed overhead capacity variance for the period?
A. $21,000 favourable
B. $21,000 adverse
C. $35,000 adverse
D. $56,000 favourable

10. Michel has the following results.

10,080 hours actually worked and paid costing $8,770

If the rate variance is $706 adverse, the efficiency variance $256 favourable, and 5,000 units were produced, what is the standard production time per unit?
A. 1.95 hours
B. 1.96 hours
C. 2.07 hours
D. 2.08 hours

Chapter 9 Time Value of Money and Capital Investment Appraisal

Learning Objectives

After the study of this chapter, you should be able to:
a) Calculate present value using compound, annuity and perpetuity formulae.
b) Explain and illustrate the payback period (PB), net present value (NPV) and internal rate of return (IRR) methods of discounted cash flow.
c) Calculate NPV, IRR and payback period (discounted and non-discounted). Interpret the results of NPV, IRR and payback calculations of investment viability

9.1 Capital Investment and Capital Expenditure Budget

9.1.1 Capital Investment

公司通過資本投資來滿足其長期資產的需要。它是公司中具有重大意義的、影響時期長的資本性支出,投資不當將對公司財務狀況產生非常不利的影響。

Companies make capital investment when they acquire non-current assets which used for a long period of time. Spending on non-current assets is capital expenditure.

Capital expenditure often represents a significant investment by a company and long term expenditure. Therefore expenditure for the wrong reasons or on the wrong assets can have a disastrous effect on a company's position.

9.1.2 Capital Expenditure and Revenue Expenditure

Capital expenditure is expenditure which results in the acquisition of non-current assets, or an improvement in their earning capacity.

Capital expenditure is not charged as an expense in the income statement of a business enterprise, although a depreciation charge will usually be made to write off the capital expenditure gradually over time. Depreciation charges are expenses in the income statement. Capital expenditure on non-current assets results in the appearance of a non-current asset in the statement of financial position of the business.

Revenue expenditure is expenditure which is incurred for

the purpose of the trade of the business or to maintain the existing earning capacity of non-current assets.

Revenue expenditure is charged to the income statement of a period, provided that it relates to the trading activity and sales of that particular period.

9.1.3 Capital Expenditure Budget

重大的投資項目必須編制項目的資本預算,進行具體的、全面的分析評價。為此在確定項目現金流量時必須區別相關成本與非相關成本、固定成本與變動成本。

The Capital expenditure budget is essentially a non-current assets purchase budget. Recurring and minor non-current asset purchases may be covered by an annual allowance provided for in the capital expenditure budget. Major projects will need to be considered individually and fully appraised. The detailed capital expenditure budget of the projects should be prepared for the budget period but additional budgets should be drawn up for both the medium and long term. In order to determine the net cash flows of investment project budget, we must identify the difference between relevant cost and non-relevant cost, fixed costs and variable costs.

(1) Relevant Cost and Non-relevant Cost

A Relevant cost is a future cash flow and arising as a direct consequence of a project's decision. It is a future, incremental cash flows as avoidable costs, differential costs, or opportunity costs etc.

A non-relevant cost is a past cash flow and is unaffected by a project's decision. It could be a sunk cost, committed cost, or notional cost (imputed cost).

Although historical costs are irrelevant for decision making, historical cost data will often provide the best available basis for predicting future costs.

(2) Fixed Costs and Variable Costs

A fixed cost is a cost which tends to be unaffected by increases or decreases in the volume of output or activity. It is a period charge as salary, rent, and depreciation.

A variable cost is a cost which tends to vary directly with the volume of output. The variable cost per unit is the same amount for each unit produced. It usually includes raw material, direct labor cost, sales commission, bonus payment. Usually variable costs will be relevant costs. General fixed costs are irrelevant to a decision, but directly attributable fixed costs are relevant costs to a decision.

Management will need to have estimates of the initial investment and future costs and revenues of a project in order to determine the net cash flows of a project and to make any long term decisions.

One of the things companies will need to consider when investing in long term projects is the time value of money.

9.2 Time Value of Money

货幣時間價值指資金隨著時間推移帶來的增額。它是項目投資評價和長期性決策必須使用的主要概念。貨幣時間價值基於三個關鍵要素：本金(P)、利率(r)和時期(n)。

Money has a time value. Think about the following question. 「If I have \$5 in my pocket now, how much will it be worth in four years' time?」Because \$5 in your pocket can be invested to earn interest or profits, the worth of \$5 in four years will be larger. The increase in amounts is the time value of money.

We will be looking at ways in which companies use the concept of the time value of money when they are appraising projects and making long term decisions.

The time value of money depends on several key factors：

a) The principal amount (P).
b) The interest rate (r).
c) The number of periods (n).

9.2.1 Interest

Money can be invested to earn interest. Hence, interest is the compensation received from giving up the right to use the cash now. Its calculation has two ways—simple interest and compound interest.

(1) Simple Interest and Compound Interest

Simple interest method is that the only principal amount accounts for interest, but interest earned do not calculate interest during the later.

The formula to calculate simple interest：
$$F = P + nPr = P \times (1+nr)$$
Where, P = The original sum invested or the principal amount
r = The interest rate (expressed as a proportion, so 10% = 0.1)
n = The number of periods (normally years)
F = The sum invested after 「n」 periods, consisting of the original capital (P) plus interest earned (future value)

In compound interest method, the interest earned is added to the principal before interest is calculated for the next year.

166　管理會計(Management Accounting)

$$F = P(1+r)^n$$

Where, P = the original sum invested
r = the interest rate (expressed as a proportion, so 10% = 0.1)
n = the number of periods (normally years)
F = the sum invested after 「n」 periods, consisting of the original capital (P) plus interests earned (future value)

Example 9.1　　Using simple interest and compound interest method Calculated separately total interests (table 9.1 and table 9.2).

Table 9.1

<table>
<tr><th colspan="5">Simple interest and compound interest
for principal amount of $10,000, at 10% over 5 years</th></tr>
<tr><th>Year</th><th>Simple interest calculation</th><th>Simple interest</th><th>Compound interest calculation</th><th>Compound interest</th></tr>
<tr><td>1</td><td></td><td></td><td></td><td></td></tr>
<tr><td>2</td><td></td><td></td><td></td><td></td></tr>
<tr><td>3</td><td></td><td></td><td></td><td></td></tr>
<tr><td>4</td><td></td><td></td><td></td><td></td></tr>
<tr><td>5</td><td></td><td></td><td></td><td></td></tr>
<tr><td></td><td>Total interest</td><td></td><td>Total interest</td><td></td></tr>
</table>

Solution:

Table 9.2

<table>
<tr><th colspan="5">Simple interest and compound interest
for principal amount of $10,000, at 10% over 5 years</th></tr>
<tr><th>Year</th><th>Simple interest calculation</th><th>Simple interest</th><th>Compound interest calculation</th><th>Compound interest</th></tr>
<tr><td>1</td><td>$10,000×10% =</td><td>$1,000</td><td>$10,000×10% =</td><td>$1,000</td></tr>
<tr><td>2</td><td>$10,000×10% =</td><td>$1,000</td><td>($10,000+1,000)×10% =</td><td>$1,100</td></tr>
<tr><td>3</td><td>$10,000×10% =</td><td>$1,000</td><td>($10,000+1,000+1,100)×10% =</td><td>$1,210</td></tr>
<tr><td>4</td><td>$10,000×10% =</td><td>$1,000</td><td>($10,000+1,000+1,100+1,210)×10% =</td><td>$1,331</td></tr>
<tr><td>5</td><td>$10,000×10% =</td><td>$1,000</td><td>($10,000+1,000+1,100+1,210+1,331)×10% =</td><td>$1,464.1</td></tr>
<tr><td></td><td>Total interest</td><td>$5,000</td><td>Total interest</td><td>$6,105.1</td></tr>
</table>

Example 9.2　　Simon invests $1,000 at 10% simple interest per annum. After one year, the original principal plus interest will amount to $1,100. How much Simon will have after 5 years?

Solution:

Chapter 9 Time Value of Money and Capital Investment Appraisal   167

The amount that Simon will have after 5 years:
F = $1,000 + (5 × 0.1 × $1,000) = $1,500

Example 9.3 Simon invests $3,000 at 10% simple interest per annum. How much Simon will have after 5 years?

Solution:
The amount that Simon will have after 5 years:
F = $3,000 + (5 × 0.1 × $3,000) = $4,500

Example 9.4 Sujata decided to invest $8,500 in an investment account paying an interest rate of 12% per annum. The interest is paid out as it is earned.
What is the amount of interest to be received at the end of each year?
If Sujata invest for 5 years, how much will be received at the end of the life of the investment?

Solution:
The amount of interest to be received at the end of each year:
I = $8,500×12% = $1,020
The amount received at the end of the life of the investment:
F = $8,500+5× $1,020 = $13,600

Example 9.5 Principle: $10,000
Interest rate: 10% per annum
Investment Period: 5 years
Calculate sum received at the end of each year.

Solution:
The sum received at the end of each year (table 9.3):

Table 9.3

Year	Principle: Opening Investment($)	Interest($)(10%)		Sum received: Closing investment($)
1	10,000	10,000×10% =	1,000	11,000
2	11,000	11,000×10% =	1,100	12,100
3	12,100	12,100×10% =	1,210	13,310
4	13,310	13,310×10% =	1,331	14,641
5	14,641	14,641×10% =	1,464.1	16,105.1

Example 9.6 $2,000 is invested for four years and interest of 8% is earned each year. What is the value of the investment at the end of year 5?

Solution:
Please look at table 9.4 as follows:

Table 9.4

Year	Principle: Opening Investment($)	Interest($)(8%)		Sum received: Closing investment($)
1	2,000	2,000×8% =	160	2,160
2	2,160	2,160×8% =	172.80	2,332.80
3	2,332.80	2,332.8×8% =	186.62	2,519.42
4	2,519.42	2,519.42×8% =	201.55	2,720.98
5	2,720.98	2,720.98×8% =	217.68	2,938.66

The value of the investment at the end of year 5:
F = $2,938.66
Or, F = $2,000 × (1+8%)^5 = $2,938.66

(2) Nominal Rate and Effective Interest Rate

利率通常按年來表述，當利息按短於一年的時間複利計算時，通常給出的年利率叫名義年利率，按實際利息計算確定的年利率叫實際年利率。

Most interest rates are expressed as per annum figures even when the interest is compounded over period of less than one year. In such cases, the given interest rate is called a nominal rate. If the interest is compounded in period less than one year, the annual percentage rate of actual interest is called an effective rate. So,

When interest is compounded once a year: Nominal rate = Effective rate

When interest is compounded in period less than 1 year (daily, weekly, monthly, quarterly or 6-monthly): Nominal rate ≠ Effective rate

Effective Interest Rate (Annual Percentage Rate, APR):
$$APR = (1 + r/n)^n - 1$$

Where APR: Annual Percentage Rate (Effective Interest Rate)
r: Interest rate per annum
n: Frequency of the interest compounded within a year.

Example 9.7   Calculate the effective annual rate of interest of:
a) 15% nominal per annum compounded quarterly.
b) 24% nominal per annum compounded monthly.

Solution:
a) The effective annual rate of interest of 15% nominal per annum compounded quarterly
$$APR = (1+15\%/4)^4 - 1 = 15.87\%$$
b) The effective annual rate of interest of 24% nominal per annum compounded quarterly
$$APR = (1+24\%/12)^{12} - 1 = 26.82\%$$

Example 9.8 Calculate annual interest and APR for each case (table 9.5):

Table 9.5

Interest compounding	Principal($)	NR(%)	Annual Interest($)	APR(%)
Yearly basis	10,000	10		
Quarterly basis	10,000	10		
Monthly basis	10,000	10		
Weekly basis	10,000	10		
Daily basis	10,000	10		

Solution:
The annual interest and APR for each case (table 9.6):

Table 9.6

Interest compounding	Principal($)	NR(%)	Annual Interest($)	APR(%)
Yearly basis	10,000	10	1,000.00	10.00
Quarterly basis	10,000	10	1,038.13	10.38
Monthly basis	10,000	10	1,047.13	10.47
Weekly basis	10,000	10	1,050.65	10.51
Daily basis	10,000	10	1,051.56	10.52

9.2.2 Compounding

複利計算是將現在的一項資金計算確定其在未來某時間的資金價值,即其未來的本利和,其結果稱終值。

Compounding is a calculation form of the time value of money. It is conversion of data from present to future and to determine future value from a known present value. A formula which can be used to show the value of an investment after several years which earns compound interest is:
$$FV = PV \times (1 + r)^n$$
Where, FV = future value of the investment after n years
PV = present value, the amount invested now
r = the rate of interest, as a proportion
n = the number of years of the investment
$(1+r)^n$ = compound factor

Example 9.9

Suppose that a project invest $5,000 now at 10%. What would the investment of the project be worth after the following number of years?

a) Five years.
b) Seven years.

Solution:
a) After five years: FV = $5,000 × (1+10%)^5
                        = $8,052.55
b) After seven years: FV = $5,000 × (1+10%)^7
                        = $9,743.59

Or, if the future value of $1 after n years at 10% interest rate is given in table 9.7 as follows:

Table 9.7   **Interest Rate**

n	Compound factor $(1+10\%)^n$
1	1.100,00
2	1.210,00
3	1.331,00
4	1.464,10
5	1.610,51
6	1.771,56
7	1.948,72

The solution is as follows:
a) After five years: FV = $5,000×1.610,51
                        = $8,052.55
b) After seven years: FV = $5,000×1.948,72
                         = $9,743.60

9.2.3 Discounting

折現計算是將未來的一項資金計算確定其在現在的資金價值,其結果稱為現值。

Discounting is another kind of calculation form of the time value of money. It is conversion of data from future to present and to determine present value from a known future value. The compound interest formula shows how we calculate a future value FV from a known current investment PV, so that if FV = PV × (1 + r)^n, then:

$$PV = FV \times \frac{1}{(1+r)^n} = FV \times (1+r)^{-n}$$

Where, FV = future value, the size of the investment after n years

PV = present value, the amount needed to invest now

r = the rate of interest, as a proportion
n = the number of years of the investment
$(1+r)^{-n}$ = discount factor

Example 9.10

A house was valued at $90,000 at the end of year 2. The price has been rising at 21% per annum. What was the value of the house at the start of the year?

Solution:
The value of the house at the start of the year was
PV = $90,000×$(1+21\%)^{-2}$ = $61,471.21

Example 9.11

An investment will generate revenue of $10,000 each year for 3 years with the first receipt at the start of year 1. Calculate the present value for the 3 years annuity based on a discount rate of 16%.

Solution(table 9.8 and table 9.9):

Table 9.8

Year	Cash flow($)	Discount factor$(1+16\%)^{-n}$	Present value($)
1	10,000	0.862,069	8,620.69
2	10,000	0.743,163	7,431.63
3	10,000	0.640,658	6,406.58
Sum	—	2.245,890	22,458.90

Or,

Table 9.9

Year	Cash flow($)	Annuity factor(16%)	Present value($)
1~3	10,000	2.245,890	22,458.90

9.2.4 Annuities

年金是間隔相等的系列等額收付款項。

Annuities are an annual cash payment or receipt which is the same amount every year for a number of years.

Future value of annuities is annuity multiply the sum of the future value factors,

$$FVA = A \times \sum_{i=0}^{n-1} (1+r)^i = A \times \frac{(1+r)^n - 1}{r} = A \times \text{Annuity} \cdot \text{future} \cdot \text{value} \cdot \text{factor}$$

According to the diagram is shown in figure 9.1 as following:

管理會計 (Management Accounting)

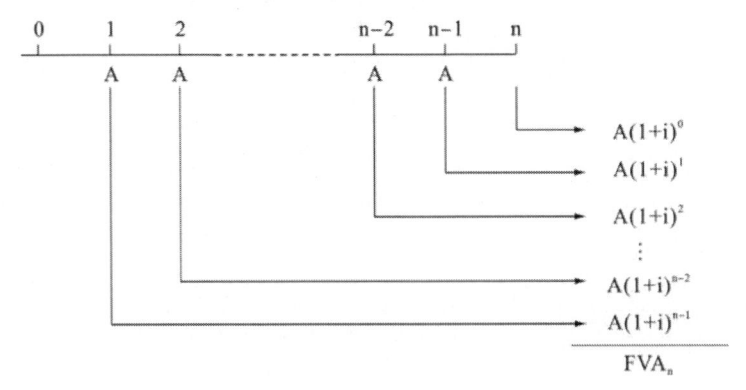

```
0   1   2        n-2  n-1   n
|   |   |         |    |    |
    A   A         A    A
                       └──→ A(1+i)⁰
                  └───────→ A(1+i)¹
              ┌───────────→ A(1+i)²
                              ⋮
              └───────────→ A(1+i)ⁿ⁻²
          └───────────────→ A(1+i)ⁿ⁻¹
                            ─────────
                               FVAₙ
```

Figure 9.1 Diagram (1)

Present value of annuities is annuity multiply the sum of the present value factors.

$$PVA = A \times \sum_{i=1}^{n} (1+r)^{-i} = A \times \frac{1 - (1+r)^{-n}}{r} = A \times \text{Annuity} \cdot \text{present} \cdot \text{value} \cdot \text{factor}$$

According to the diagram is shown in figure 9.2 as following:

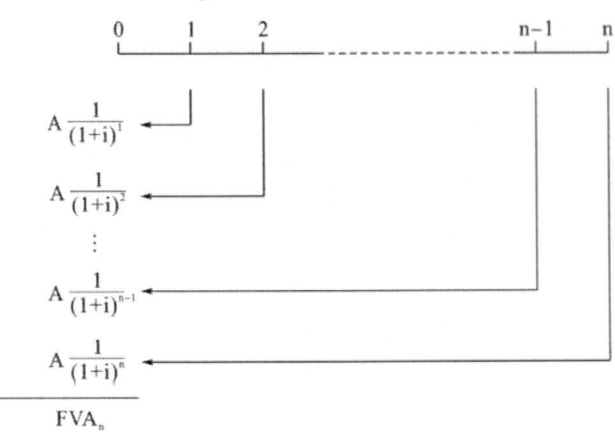

```
       0   1   2              n-1   n
       |   |   |               |    |
   A·1/(1+i)¹ ←┘
   A·1/(1+i)² ←────┘
        ⋮
   A·1/(1+i)ⁿ⁻¹ ←──────────────┘
   A·1/(1+i)ⁿ  ←───────────────────┘
   ─────────
      FVAₙ
```

Figure 9.2 Diagram (2)

Example 9.12

A project would involve a capital outlay of $85,000. Profits (before depreciation) of the project would be $21,000 each year. The cost of capital is 12%. Would the project be worthwhile if it lasts:

a) five years.
b) eight years.

Solution:

a) If the project lasts five years, please look at table 9.10 as following:

Table 9. 10

Year	Cash flow($)	Annuity factor(12%)	Present value($)
0	-85,000	1	-85,000
1~5	21,000	3.604,8	75,700.30
Sum			-9,299.70

b) If the project lasts eight years, please look at table 9.11 as following:

Table 9. 11

Year	Cash flow($)	Annuity factor(12%)	Present value($)
0	-85,000	1	-85,000
1~8	21,000	4.967,6	104,320.44
Sum			19,320.44

The project is not worthwhile if it lasts only five years, but it would be worthwhile if it lasts eight years.

Example 9. 13 A project invests $38,500. It would earn $12,000 each year for the first three years and then $8,000 each year for the next three years. Cost of capital is 12%. Is the project worth undertaking?

Solution:

Present value of $1 per annum, year 1~3   2.401,8
Present value of $1 per annum, year 1~6   4.111,4
Present value of $1 per annum, year 4~6 (=4.111,4-2.401,8)   1.709,6

Table 9. 12

Year	Cash flow($)	Annuity factor(12%)	Present value($)
0	-38,500	1	-38,500
1~3	12,000	2.401,8	28,821.98
4~6	8,000	1.709,6	13,676.61
Sum			3,998.58

The sum present value of the project is positive, so it worth undertaking.

9.2.5 Perpetuities

永續年金是沒有最終期限的年金,它只有現值、沒有終值。

Perpetuities are annuities which are expected to continue for an indefinitely long period of time. It only has present value, there is no final value.

PV of perpetuity = Annuity ÷ Interest rate
                 = A / r

Example 9.14 What is the present value of a perpetual annuity of $500 p.a. at 10%?

Solution:
PV of perpetuity = $500÷10% = $5,000

9.3 Using Non-discounted Cash Flow Models to Make Capital Investment Appraisal

In this topic, we will see how companies use capital investment analysis techniques to decide which long-term capital investments to make. In the first place, we see how to use the payback period and ARR models to make capital investment appraisal.

9.3.1 Payback period (PB)

投資回收期是用投資項目實現的現金淨流量收回初始投資所需要的時間。

The payback period is the time taken for the initial investment to be recovered in the net cash inflows from the project.

The payback method is particularly relevant if there are liquidity problems or distant forecasts are very uncertain. The payback method measures how quickly managers expect to recover their initial investment. Usually, the shorter the payback period of the project is, the more attractive the asset. So decision rule of payback period is investments with shorter payback periods are more desirable, all else being equal.

The payback period is the length of time required before the total net cash inflows from the project to equal the original cash outflows. Calculating the payback period of the project depends on whether net cash inflows are equal each year, or whether they differ over time.

(1) Payback Period with Constant Annual Net Cash Inflows

If the expected net cash inflows from a project are an equal annual amount, the payback period is calculated simply as:

$$\text{Payback} = \frac{\text{Initial payment}}{\text{Expected annual net cash inflow}}$$

It is normally assumed that net cash flows each year occur at an even rate throughout the year.

Example 9. 15   An expenditure of $3 million is expected to generate cash inflows of $800,000 each year for the next seven years. What is the payback period for the project?

Solution:
Payback period for the project
= $3,000,000 ÷ $800,000 = 3.75 years

Example 9. 16   A project will involve spending $2.8 million now. Annual cash flows from the project would be $500,000.
What is the expected payback period?

Solution:
The expected payback period = $2,800,000 ÷ $500,000 = 5.6 years

(2) Payback Period with Unequal Annual Net Cash Inflows

Annual net cash flows from a project are unlikely to be a constant annual amount, but are likely to vary from year to year.

The simplest way of calculating payback is probably to set out the figures in a table. The payback period is calculated as shown in example 9.17, example 9.18.

Example 9. 17   A project is expected to have the following net cash flows (table 9.13).

Table 9. 13   Net Cash Flow (1)

Year	Net Cash Flow ($'000)
0	(2,000)
1	900
2	850
3	700
4	500

What is the expected payback period?

Solution:
When periodic net cash flows are unequal, you must total net cash inflows until the amount invested is recovered. So the payback period is calculated as below (table 9.14):

Table 9.14　　　　　　　　Calculative Net Cash Folw (1)

Year	Net Cash Flow($'000)	Cumulative net cash flow($'000)
0	(2,000)	(2,000)
1	900	(1,100)
2	850	(250)
3	700	450
4	500	950

The expected payback period
= 2 years + $250(amount needed to complete recovery in year 3) ÷ $700(net cash inflow in year 3)
= 2.36 years

Example 9.18　　Calculate the payback period in years and months for the following project(table 9.15).

Table 9.15　　　Net Cash Flow (2)

Year	Net Cash Flow($'000)
0	(3,100)
1	2,200
2	800
3	700
4	600
5	600
5-Residual Value	900

Solution:
The payback period is calculated as below (table 9.16):

Table 9.16　　　　　　　Cumulative Net Cash Flow (2)

Year	Net Cash Flow($'000)	Cumulative cash flow($'000)
0	(3,100)	(3,100)
1	2,200	(900)
2	800	(100)
3	700	600
4	600	1,200
5	600	1,800
5-Residual Value　900		—

The payback period
= 2 years + $100(amount needed to complete recov-

ery in year 3)/ \$700(net cash inflow in year 3)
= 2.14 years

A major criticism of the payback method is that it focuses only on time, not on profitability. The payback period considers only those net cash flows that occur during the payback period. It ignores any net cash flows that occur after that period.

9.3.2 Accounting rate of return (ARR)

會計收益率反應投資項目盈利能力，它關注投資項目實現的經營收益，而非現金流量。會計收益率等於項目平均年經營收益除以項目的平均投資額，體現投資項目在整個投資期內的平均收益水平。

Accounting rate of return (ARR) on a project is a measure of profitability of the invested project. The ARR focuses on the operating income, not the net cash inflow, a project generates. The ARR measures the average rate of return over the project's entire life. If company that use the ARR model set a minimum required accounting rate of return, then managers would make the decision whether invest a project. So decision rule of ARR is that if the expected ARR of a project exceeds the required rate of return, managers would invest the project, if the expected ARR of a project is less than the required rate of return, managers would not invest the on project.

Accounting rate of return on a project is calculated simply as below:

$$\text{Accounting rate of return} = \frac{\text{Average annual operating income from a project}}{\text{Average amount invested in a project}}$$

Total operating income during operating life of a project is total net cash inflows during operating life of the project less total depreciation during operating life of the project (equal the initial investment less residual value of the project), then average annual operating income from the project is total operating income during operating life of a project divide by the project's operating life in years. And average amount invested in the project is the initial investment of the project plus the residual value of the project at the end of the project's operating life divided by two.

Example 9.19

Same data as example 9.17, calculate accounting rate of return on the project.

Solution:
Total operating income during operating life
= (900+850+700+500)-(2,000-0)= 950

Average annual operating income
= 950÷4 = 237.5
Average amount invested = (2,000+0)/2 = 1,000
So, Accounting rate of return (ARR)
= 237.5÷1,000 = 23.75%

Example 9.20 Same data as example 9.18, calculate accounting rate of return on the project.

Solution:
Total operating income during operating life
= (2,200+800+700+600+600) − (3,100−900) = 2,700
Average annual operating income
= 2,700÷5 = 540
Average amount invested = (3,100+900)÷2 = 2,000
So, Accounting rate of return (ARR)
= 540÷2,000 = 27%

In summary, the payback period focuses on the time it takes for the company to recoup its cash investment but ignores all cash flows occurring after the payback period. Because it ignores any additional cash flows (including any residual value), the method does not consider the profitability of the project.

The ARR, however, measures the profitability of the project over its entire life using accrual accounting figures. It is the only method that uses accrual accounting rather than net cash inflows in its calculations.

The payback period and ARR methods are simple and quick to calculate, so managers often use them to screen out undesirable investments. However, both methods ignore the time value of money.

9.4 Using Discounted Cash Flow Models to Make Capital Investment Appraisal

Either the payback period or the ARR methods ignores the time value of money. That is, these methods fail to consider the timing of the net cash inflows a project generates. Discounted cash flow (DCF) methods can overcome this weakness. Discounted cash flow method is a technique of evaluating capital investment projects, using discounting arithme-

Chapter 9　Time Value of Money and Capital Investment Appraisal　179

tic to determine whether or not they will provide a satisfactory return. Discounted cash flow techniques take account of the time value of money. These methods incorporate compound interest by assuming that companies will reinvest future cash flows when they are received. Many companies that manufacture products or provide services use discounted cash flow methods to make capital investment decisions. The key discounted cash flow methods are:
a) Net present value (NPV) method.
b) Internal rate of return method.
c) Discounted payback (D.PB) method.

9.4.1 Net Present Value (NPV)

淨現值是投資項目所有現金淨流量現值的代數和,也是投資項目所有現金淨流入量的現值減去所有現金淨流出量的現值的差額。若投資項目的淨現值大於 0,投資項目能增加企業價值,因此投資項目可接受,具有財務可行性;若投資項目的淨現值小於 0,投資項目會降低企業價值,因此投資項目不能接受,不具有財務可行性。淨現值越大,投資項目越好。

The net present value method calculates the present value of all cash flows, and sums them to give the net present value. In other word, the net present value is the net difference between the present values (PV) of the investment's cash inflows and the investment's cash outflows (the initial investment). So:

If PV of cash inflow > PV of cash outflow = + NPV = Profits
If PV of cash outflow > PV of cash inflow = - NPV = Loss

Therefore, NPV indicates the immediate increase or decrease in wealth resulting from the acceptance of the project. So decision rule of NPV is that if NPV is positive, then the project is acceptable, if NPV is negative, the project would not be worth investing in.

The net present value on a project is calculated simply as below:
NPV = PV of cash inflow - PV of cash outflow

Example 9.21

Investment AB requires an initial investment of \$150,000. It is expected to generate annual cash inflow of \$30,000 for 5 consecutive years.
The cost of capital (or the rate of return) is 10%.
Determine the net present value for the investment.

Solution:
The net present value for the investment
　　= \$30,000×(P/A, 10%, 5) - \$150,000
　　= - \$36,276.4

Example 9.22   Investment AB requires an initial investment of $150,000. It is expected to generate annual cash flow in table 9.17 as following:

Table 9.17   Annual Cash Flow

Year 1	Year 2	Year 3	Year 4	Year 5
$30,000	$40,000	$60,000	$80,000	($20,000)

The cost of capital (or the rate of return) is 10%.
Determine the net present value for the investment.

Solution:
The net present value for the investment is calculated in table 9.18 as following:

Table 9.18   The Net Present Value for the Investment

Year	New cash flow ($)	Discount fator $1/(1+10\%)^n$	Present value ($)
0	-150,000	1	-150,000
1	30,000	0.909,1	27,272.73
2	40,000	0.826,4	33,057.85
3	60,000	0.751,3	45,078.89
4	80,000	0.683,0	54,641.08
5	-20,000	0.620,9	-12,418.43
NPV of investment			-2,368

Example 9.23   An investment has the following expected cash flows over its economic life of three years, please look at table 9.19:

Table 9.19   Cash Flow of Three Years

Year	Net cash flow ($)
0	(142,700)
1	51,000
2	62,000
3	73,000

Calculate the net present value (NPV) of the project at discount rates of 0%, 10% and 20%.

Solution:
The net present value of the project is calculated in table 9.20 as below:

Table 9.20   The Net Present Value of the Project

Year	New cash flow ($)	Discount fator 1/(1+0%)ⁿ	Present value ($)	Discount fator 1/(1+10%)ⁿ	Present value ($)	Discount fator 1/(1+20%)ⁿ	Present value ($)
0	−142,700	1	−142,700	1	−142,700	1	−142,700
1	51,000	1	51,000	0.909,1	46,363.64	0.833,3	42,500.00
2	62,000	1	62,000	0.826,4	51,239.67	0.694,4	43,055.56
3	73,000	1	73,000	0.751,3	54,845.98	0.578,7	42,245.37
NPV of investment	—	43,300	—	9,749.29	—	−14,899.07

9.4.2 Internal Rate of Return (IRR)

Internal rate of return (IRR) is the rate of return (based on discounted cash flows) a company can expect to earn by investing in the project. It is the discount rate that makes the net present value (NPV) of the investment's project equal to zero.

The net present value (NPV) on a project is calculated by:
NPV = PV of cash inflow−PV of cash outflow
If NPV=0, then
PV of cash inflow=PV of cash outflow

內含報酬率是投資項目淨現值等於 0 時的折現率,反應投資項目內在的收益水平。若投資項目的內含報酬率大於企業要求的投資收益率,投資項目可以接受;若內含報酬率小於企業要求的投資收益率,投資項目不能接受。內含報酬率越高,投資項目越好。

In other words, the Internal rate of return (IRR) is the discount rate that makes the present value of the investment's net cash inflows equal to the present value of the investment's net cash outflows (the cost of the investment). So decision rule of IRR is that if the IRR of a project exceeds the required rate of return, the project is acceptable, if the IRR of a project is less than the required rate of return, the project is unacceptable. The higher the IRR, the project is the more desirable.

Calculating the internal rate of return (IRR) of the project depends on whether net cash inflows are equal each year, or whether they differ over time.

(1) Internal Rate of Return (IRR) with Constant Annual Net Cash Inflows

The IRR is discount rate at which the NPV will become zero (Similar to the breakeven point for sales revenue). In other words, the IRR is the discount rate that makes the present value of the investment's net cash inflows equal to the present value of the cost of the investment. So we calculate the IRR of a project with constant annual net cash inflows (annuity) by taking the following steps:

a) Setting up the following equation:
Project's cost = Amount of each equal net cash inflow × Annuity PV factor (i, n)
b) Then rearrange the equation and solve for the Annuity PV factor (i, n):
Annuity PV factor (i, n) = Project's cost / Amount of each equal net cash inflow
c) Finally, find the interest rate that corresponds to this Annuity PV factor using the Present Value of Annuity of $1 table (Appendix, Table C). In a project, if the exact Annuity PV factor (i, n) appears in the Present Value of Annuity of $1 table (Appendix, Table C), the IRR of the project equal to i. Many times, the exact Annuity PV factor (i, n) will not appear in the table. In this case, we may find the closest two factors and interest rates, then calculate the approximation of the IRR of the project showing in example 9.25.

Example 9.24

An investment of $3,049,200 for a project is expected to generate cash inflows of $800,000 each year for the next seven years.
What is the internal rate of return (IRR) for the project?

Solution:
$3,049,200 = $800,000 × Annuity PV factor (i, 7)
Annuity PV factor (i, 7) = $3,049,200 / $800,000
= 3.811,5
Finding factor(= 3.811,5) using the Present Value of Annuity of $1 table (Appendix, Table C), then
Annuity PV factor (18%, 7) = 3.811,5, So:
IRR = 18%

Example 9.25

A project will involve spending $2.3 million now. Annual cash flows from the project would be $500,000 for the next ten years.
What is the expected internal rate of return (IRR) for the project?

Solution:
$2,300,000 = $500,000 × Annuity PV factor (i, 10)
Annuity PV factor (i, 10) = $2,300,000 ÷ $500,000 = 4.5
Finding factor(= 4.5) using the Present Value of Annuity of $1 table (Appendix, Table C), then find out the closest

two factors and interest rates as below:
Annuity PV factor (14%, 10) = 5.216,1
Annuity PV factor (i, 10) = 4.5
Annuity PV factor (16%, 10) = 4.833,2
Assume that the factors and interest rates proportional relationship changes as following figure 9.3.

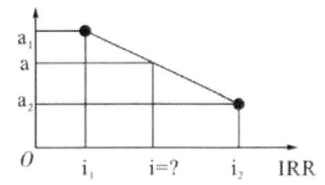

Figure 9.3 The Relationship Between the Factors and IRP

then,
$$\frac{(i-15\%)}{(16\%-15\%)} = \frac{(4.5-5.216,1)}{(4.833,2-5.216,1)}$$

$$i = 14\% + \frac{(4.5-5.216,1)}{(4.833,2-5.216,1)} \times (16\%-14\%) = 17.74\%$$

So, the expected internal rate of return (IRR) for the project:
IRR = i = 17.74%

(2) Internal Rate of Return (IRR) with Unequal Annual Net Cash Inflows

When a project has unequal cash inflows, we cannot use the Present Value of Annuity of $1 table to find the project's IRR. Rather, we must use a trial-and error procedure to determine the discount rate making the project's NPV equal to zero. For example, if the company's minimum required rate of return is 10%. Recall from example 9.23 that the project's NPV using a 10% discount rate is $9,749.29. Since the NPV is positive, the IRR must be higher than 10%. We continues the trial-and error process using higher discount rates until the company finds the rate that brings the net present value of the project to zero. The consequence in the following table shows that at 12%, the project has an NPV of $4,221.69. Therefore, the IRR must be higher than 12%. At 14%, the project's NPV is $-983.25, which is negative, but very close to zero. Thus, the IRR must be slightly lower than 14%, and be calculate the approximation as below in table 9.21:

$$\frac{(IRR-12\%)}{(14\%-12\%)} = \frac{(0-4,221.69)}{(-983.25-4,221.69)}$$

$$IRR = 12\% + \frac{(0-4,221.69)}{(-983.25-4,221.69)} \times (14\%-12\%) = 13.62\%$$

Table 9.21　　　　　　　　　　**Internal Rate of Return**

Year	New cash flow ($)	Discount fator $1/(1+10\%)^n$	Present value ($)	Discount fator $1/(1+12\%)^n$	Present value ($)	Discount fator $1/(1+14\%)^n$	Present value ($)
0	-142,700	1	-142,700	1	-142,700	1	-142,700
1	51,000	0.909,1	46,363.64	0.892,9	45,535.71	0.877,2	44,736.84
2	62,000	0.826,4	51,239.67	0.797,2	49,426.02	0.769,5	47,706.99
3	73,000	0.751,3	54,845.98	0.711,8	51,959.96	0.675,0	49,272.92
NPV of investment		—	9,749.29	—	4,221.69	—	-983.25

Example 9.26

The company's minimum required rate of return is 16%. An investment has the following expected cash flows over its economic life of five years, please look at table 9.22 as below:

Table 9.22　　　　**The Cash Flows**

Year	$
0	-256,800
1	65,000
2	73,900
3	82,400
4	97,600
5	115,100

Calculate the Internal Rate of Return (IRR) of the investment's project, judge whether to accept the project.

Solution:
Using a trial-and error procedure, show the following table 9.23:

Table 9.23

Year	New cash flow ($)	Discount fator $1/(1+16\%)^n$	Present value ($)	Discount fator $1/(1+18\%)^n$	Present value ($)	Discount fator $1/(1+20\%)^n$	Present value ($)
0	-256,800	1.000,0	-256,800	1.000,0	-256,800	1.000,0	-256,800
1	65,000	0.862,1	56,034.48	0.847,5	55,084.75	0.833,3	54,166.67
2	73,900	0.743,2	54,919.74	0.718,2	53,073.83	0.694,4	51,319.44
3	82,400	0.640,7	52,790.19	0.608,6	50,151.18	0.578,7	47,685.19
4	97,600	0.552,3	53,903.61	0.515,8	50,340.99	0.482,3	47,067.90
5	115,100	0.476,1	54,800.61	0.437,1	50,311.27	0.401,9	46,256.11
NPV of investment		——	15,648.63	——	2,162.02	——	-10,304.69

$$\frac{(IRR-18\%)}{(20\%-18\%)}=\frac{(0-2,162.02)}{(-10,304.69-2,162.02)}$$

$$IRR=18\%+\frac{(0-2,162.02)}{(-10,304.69-2,162.02)}\times(20\%-18\%)=18.35\%$$

The Internal Rate of Return (IRR) of the investment is 18.35%, higher than 16% required rate of return, so the investment's project would be accepted.

9.4.3 Discounted Payback Period (D.PB)

The discounted payback technique is an adaptation of the payback technique, which we looked at earlier, taking some account of the value of money. The Discounted payback (D.PB) method applies discounting to arrive at a payback period after which the NPV become positive. The discounted payback period is the time it will take before a project's cumulative NPV turns from being negative to being positive.

Example 9.27

If we have a cost of capital of 10% and a project with the cash flow shown in table 9.24 below, calculate the discounted payback period using the table 9.25.

Table 9.24   The Cash Flow

Year	Cash Flow ($)	Discount factor	Present Value ($)	Cumulative NPV ($)
0	-100,000			
1	30,000			
2	50,000			
3	40,000			
4	30,000			
5	20,000			

Solution:

Table 9.25   The Calculation of Discounted Payback Period

Year	Cash Flow ($)	Discount factor $1/(1+10\%)^n$	Present Value ($)	Cumulative NPV ($)
0	-100,000	1.000,0	-100,000	-100,000
1	30,000	0.909,1	27,272.73	-72,727.27
2	50,000	0.826,4	41,322.31	-31,404.96
3	40,000	0.751,3	30,052.59	-1,352.37
4	30,000	0.683,0	20,490.40	19,138.04
5	20,000	0.620,9	12,418.43	31,556.46

The discounted payback period
= 3+1,352.37÷20,490.40=3.07 years

9.5 Comparing Capital Investment Appraisal Models

We have discussed five capital investment appraisal models commonly used by companies to make capital investment decisions. There are two methods that ignore the time value of money: payback period and accounting rate of return. In comparison, they have the following characteristics.

(1) Payback Period Method
a) Simple to compute.
b) Focuses on the time it takes to cover the company's cash investment.
c) Ignores any cash flows occurring after the payback period.
d) Highlights risks of investments with longer cash recovery periods.
e) Ignores the time value of money.

(2) Accounting Rate of Return Method
a) The only method that uses accrual accounting figures.
b) Shows how the investment will affect operating income, which is important to financial statement users.
c) Measures the profitability of the project over its entire life.
d) Ignores the time value of money.

There are three methods that incorporate the time value of money: net present value, internal rate of return, discounted payback.

(3) Net Present Value Method
a) Considers the time value of money and the project's net cash flows over its entire life.
b) Indicates whether or not the project will earn the company's minimum required rate of return.
c) Shows the excess or deficiency of the project's present value of net cash inflows over its initial investment cost. If a project with a positive NPV is accepted, company shareholders will benefit and shareholder wealth is maximized.

d) The profitability index should be calculated for capital rationing decisions when the project requires different initial investments.

e) Need to make sure an appropriate discount rate for the project but it is difficult.

(4) Internal Rate of Return Method

a) Considers the time value of money and the project's net cash flows over its entire life.

b) Calculates the project's unique rate of return.

c) No additional steps needed for capital rationing decisions.

d) Don't need to make sure an appropriate discount rate for the project.

The discounted payback (D.PB) method, like the basic payback method, fails to take account of positive cash flows occurring after the end of the payback period etc. But it takes into account the time value of money.

QUESTIONS:

1. A company has evaluated the potential for investment in Project Y.
Cash flow (table 9.26) is estimated to be:

Table 9. 26　　　　　　　　　　The Cash Flow

Year	0	1	2	3	4	5
Project Y ($'000)	(1,000)	400	400	400	400	400

(i) Calculate the payback period for the project.

(ii) Calculate the net present value (NPV) of the project using a discount rate of 12% per annum.

(iii) Determine the internal rate of return (IRR) of the project.

2. A company has to choose between two projects, Project A and Project B. Cash inflow from projections are as shown in table 9.27:

Table 9. 27　　　　　　　　　Cash Inflow Projections

Year	Project A ($000)	Project B ($000)
1	100	200
2	200	200
3	300	200
4	400	200

Year	Project A ($000)	Project B ($000)
5	500	200

The projects require an initial investment of:
Project A $900,000
Project B $640,000
a) Calculate for each project:
(i) The net present value.
(ii) The internal rate of return.
b) Calculate the payback period.

3. N Company has been looking at a potential project which has the following cash inflows (table 9.28):

Table 9.28　　　　　　　　　　The Cash Inflows

End of year	$'000
1	15
2	17
3	22
4	2

To acquire these inflows N Company would have to invest $36,000 in fixed assets now. The assets would be expected to be sold at the end of the project for $2,000.
a) Calculate the net present value of the project using 10% and 20% as the discount factors.
b) Using your results from part (a), calculate the internal rate of return for this investment to one decimal place.

4. A company has decided to lease a machine. Six annual payments of $8,000 will be made with the first payment on receipt of the machine. Below is an extract from an annuity table (table 9.29).
What is the present value of the lease payments at an interest rate of 10%?
A. $30,328　　　　　　　　　B. $34,840
C. $38,328　　　　　　　　　D. $48,000

Table 9.29　　　　　　　　　　Annuity Table

Year	Annuity factor (10%)
1	0.909,1
2	1.735,5
3	2.486,9
4	3.169,9
5	3.790,8
6	4.355,3

5. An investment project has a positive net present value (NPV) of $7,222 when its cash flows are discounted at the cost of capital of 10% per annum. Net cash inflows from the project are expected to be $18,000 per annum for five years. The cumulative discount (annuity) factor for five years at 10% is 3.791.
What is the investment at the start of the project?
A. $61,016   B. $68,238
C. $75,460   D. $82,778

6. What is the effective annual rate of interest of 2.1% compounded every three months?
A. 6.43%   B. 8.40%
C. 8.67%   D. 0.87%

7. A company is considering an immediate investment in new machinery. The machinery would cost $100,000 with expected net cash inflows of $30,000 per year starting in year 1. The disposal value of the machine after five years is expected to be $10,000. $15,000 has already been incurred on development costs.
What is the payback period of the investment based on future incremental cash flows?
A. 3.0 years   B. 3.3 years
C. 3.5 years   D. 3.8 years

8. Net cash flows, estimated for a capital investment project, have been discounted at four discount rates with the following results (table 9.30):

Table 9.30   **Discount Rate and Net Present Value**

Discount rate	5%	10%	15%	20%
Net present value ($000)	92.9	39.1	(4.8)	(40.9)

What is the best estimate of the IRR using only the above data as appropriate?
A. 13.6%   B. 14.5%
C. 15.4%   D. 15.7%

9. A project, investing in new machinery, has an estimated five year life. The cost of capital is 10% per annum.
Estimated cash flows are shown in table 9.31 as below:

Table 9.31   **Estimated Cash Flows**

Time	Cash flows
0 (cost)	($186,000)
1 to 5 (inflows)	$56,000 per annum
5 (residual value)	$10,000

The cumulative discount factor at 10% for time 1 to 5 is 3.79. The discount factor at 10% for time 5 is 0.62.

What is the net present value of the project?
A. $16,240
B. $20,040
C. $32,440
D. $36,240

Chapter10 Performance Measurement

Learning Objectives

After the study of this chapter, you should be able to:
a) Understand how the purpose and strategy of organization relate to performance measurement.
b) Explain the meaning of, and calculate, controllable profits, ROI and RI in measuring divisional performance.
c) Explain why and how to use multiple measures—financial and non-financial measures.
d) Understand the concept of critical success factor and key performance Indicator.
e) Describe the balance scorecard and its four perspectives.

10.1 Performance Measurement Overview

業績管理是對員工、部門、企業或某個系統或某要素的相關業績訊息進行收集、分析、報告的過程。

Performance measurement is the process of collecting, analyzing and/or reporting information regarding the performance of an individual, department, organization, system or component.

The results of performance management exercises is used:
a) As factual basis for remunerations and rewards.
b) In employee development training and promotion.
c) To direct and support individual/department so that they can work as efficiently and effectively as possible.
d) In monitoring to make sure the business on the track.
e) For making informed decisions on business future priorities and resource allocation.

Performance measurement is an integral part of management control system. Effectively managing an organization's performance should be tied to the strategy objectives of the organization(figure 10.1). It shall begin with a clear understanding of what the organization is trying to achieve, what's being done today, what progress is being made, and what's needed to ensure the organization's goals being met.

```
strategic objectives
     and goals
         ⇩
  critical success
       factor
         ⇩
  key performance
      indicator
```

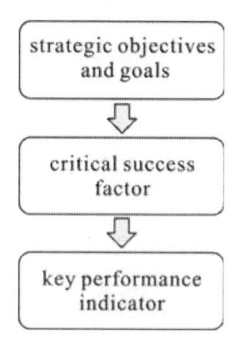

Figure 10.1　Effective Performance Measurement

10.1.1　What the Organization Is Trying to Achieve

首先,明確企業的願景和長期戰略,然後確定中長期目標。

「You measure what you value, you get what you measure.」

Before implement performance evaluation, we need to clarify the purpose and objectives of the organization. Why does the organization exist? What is the long term strategy for the organization? What is crucial for business success? Then from the vision and strategy, we derive goals and objectives.

Objectives concern organization as a whole, for example, please look at table 10.1 as following:

Table 10.1　　Organization Objectives

a) Profitability	e) Customer satisfaction
b) Market share	f) Quality
c) Growth	g) Earnings per share
d) Risk	h) Added value

After strategic and operational goals were set, the company has to determine how to measure whether objectives and goals have been met and how to achieve them effectively.

10.1.2　How to Measure Whether the Organization Achieves Its Goal

其次,設定一系列業績評價指標。

Performance measures, or indicators, are developed as standards for assessing the extent to which these objectives are achieved. For example, if a software company's goal is to have the fastest growth in its industry, its main performance indicator may be the measure of revenue growth year-on-year.

10.1.3　Three E's to Evaluate Public Sector Performance

Economy, efficiency and effectiveness are all generally desirable features of the performance of public sectors, such as hospitals and schools.

Economy represents value for money and lies in operating at

Chapter10 Performance Measurement

經濟、高效、有效是衡量公共服務單位業績的標準。

minimum cost. However, an over-parsimonious approach will reduce effectiveness.

Effectiveness is achieving established objectives. There are usually several ways to achieve objectives, some more costly than others.

Normally quantifiable indicators are used to measure effectiveness, such as exam result, average class size in school, crime rate, hospital waiting time.

Efficiency consists of attaining desired results at minimum cost. It therefore combines effectiveness with economy.

To assess the efficiency of public service provided, financial indicators, for example, cost per bed-night in a hospital, administration costs as a proportion of total cost in a college, are applicable.

10.2 Application of Performance Measurement for Organization

10.2.1 Critical Success Factor

找出企業競爭力的關鍵成功因素。

Performance measure plays an important part of the control process. The essence of control is that actual performance is compared with a standard or target that was established earlier. Modern technology is capable of producing massive management accounting information, but not all of them are key to organization's success. Manager should be able to identify the critical success factor that is fundamental to competitive success.

Critical success factor[1] can be set and used by identifying objectives and goals, determining which factors are critical for accomplishing each objective and then determining a small number of performance measures for each factor. CSF may include items such as product, quality, employee attitudes, manufacturing flexibility and brand awareness (CIMA[2] official terminology).

CSFs vary from industry to industry and even time to time within the same industry as driving forces and competitive conditions change.

There are some examples of CSFs that have been identified for contemporary major industries:

---

[1] CSF had been first introduced by D Ronald Daniel, later managing director of the management consultancy McKinsey and Co in 1962.

[2] CIMA (The Chartered Institute of Management Accountants), 英國特許管理會計師公會, 全球最大的國際性管理會計師組織。

a) In the food processing industry, new product development, good distribution channels, effective advertising.

b) In the supermarket industry, the right product mix available in each store, having it actually available on the shelves, advertising it effectively to pull shoppers in, pricing it correctly (since profit margins were low in this industry).

c) In mobile and computer industry, the reliability of technology, user experience and new innovative features

d) In the automobile industry, the reliability of technology, after sale service, safety and fuel efficiency.

10.2.2 Key Performance Indicators (KPI)

主要業績指標(KPI): 衡量企業戰略目標是否實現以及各部門單位、各員工表現的一系列量化指標。

Once the critical areas have been identified, the organization should be able to measure its performance in those areas normally by using key performance indicator (KPI), which is a set of quantifiable measures that a company uses to review how successfully a strategy has been implemented and how well an organization is performing.

Some examples of KPI are sales, customer satisfaction, percentage of returning customers, percentage of defective product declined by inspection, share price (for public company, which might be the best KPI for measuring how well the management maximized shareholders value), and revenue growth (especially for early-stage company).

KPIs give everyone in the organization a clear picture of what is important, of what they need to make happen. KPI works as not only a management tool but also to motivate employees to achieve or exceed the KPIs.

In selecting key performance indicators, it is critical to make sure them follow the SMART criteria:

a) Specific — the measure has a specific purpose for the business and clearly defined.

b) Measurable — the KPI must have a value.

c) Attainable — the measure have to be achievable.

d) Relevant — relevant to the success of the organization.

e) Time bound — the value or outcome are shown for a predefined and relevant period.

10.2.3 Critical Success Factor vs Key Performance Indicator

Critical success factors are elements that are vital for a strategy to be successful.

A critical success factor drives the strategy forward; It

makes or breaks the success of the strategy (hence 「critical」). Strategists should ask themselves 「Why would customers choose us」. The answer is typically a critical success factor.

KPIs, on the other hand, are measures that quantify management objectives, along with a target or threshold, and enable the measurement of strategic performance. An example:

KPI = Number of new customers. (Measurable, quantifiable) + Threshold = 10 per week (KPI reached if 10 or more new customers, failed if <10)

CSF = Installation of a call centre for providing superior customer service (and indirectly, influencing acquiring new customers through customer satisfaction).

10.2.4 Financial Performance Measures

財務指標是滯後的,是經營活動的結果。

Financial performance is fundamental to business. It is not surprising that financial indicators are the most commonly used KPI across all industry and sectors.

These measures include profit, revenue, costs, share price, earnings per share, cash flow, and so on.

Moreover, we calculate financial ratio to evaluate how well the organization perform in aspect of liquidity and solvency, operating efficiency and profitability.

Liquidity and solvency: Quick ratio, acid test ratio, long-term debt to asset ratio, etc;
Operating efficiency: asset turnover, inventory turnover, receivable turnover, etc;
Profitability: return on asset, return on investment, etc.

These financial results should be compared with the following: Budgeted sales, costs and profits / results of other parts of the business / results of other business / the economy in general.

However, only financial performance measure is inadequate to assess whether the organization achieves its goal. In fact, financial results which are considered as lagging indicators are the consequence, not the causes of a series of operational activities and process. In order to be successful, organizations have to perform well across a range of key process before financial results settle. Therefore, CSF and KPI should not only focus on financial performance but also non-financial performance such as product, customers, after-sale service and even employees moral.

10.2.5 Non-financial Performance Measure

In recent year, due to the change of business environment, non-financial indicator draws more and more attention as they are considered to be leading indicators of financial performance. The use of non-financial performance measure has significantly increased.

(1) Changes in Manufacturing Environment

Organizations are increasingly using quantitative and qualitative non-financial indicators, instead of focusing on costs reduction only.

非財務指標是前瞻指標。

(2) Changes in Competitive Environment

The businesses are competing in terms of product quality, delivery, reliability, after-sale service and customer satisfaction. None of these variables is directly measured by traditional financial system.

For example, the manufacturing department may have a KPI of 「number of units rejected by quality inspection」, while the sales department has a KPI of 「minutes a customer is on hold before a sales rep answers」. Success by the sales and manufacturing departments in meeting their respective departmental key performance indicators will help the company meet its overall KPI.

One of the approaches to integrate non-financial indicators into performance measurement is the balance scorecard, which we will discuss later in this chapter.

Practice

For a delivery company, make up five financial indicators and five non-financial indicators.

10.3 Application of Performance Measurement for Responsibility Centers

In order to ensure that junior managers in an organization make decisions that are in the best interests of the organization as a whole, senior managers generally introduce the following system of performance measures, please look at table 10.2.

Table 10.2   The System of Performance Measures

Responsibility center	Manager responsible for	Financial performance measure
Cost center	Costs	Variances
Revenue center	Revenues	Revenues
Profit center	Costs and revenues	Controllable profit
Investment center	Costs, revenues and assets	ROI and RI

10.3.1 Measuring Performance-Cost centers

As we discussed in early chapter, variance analysis is often used as a way of measuring the performance of cost centers. Be aware that:

a) The variances are only as good as the standards on which they are based.
b) It is often difficult to identify which costs are controllable and which are not.

10.3.2 Measuring Performance-Profit Centers

Controllable profit/contribution is the most appropriate performance measure of a profit center. Thus, it is important to distinguish between controllable and non-controllable costs. For example, depreciation on division machinery, head office finance and legal staff who are assigned to providing services for specific division, which category do they fall into? Sometimes the answer is not as obvious as we think.

10.3.3 Measuring Performance-Investment Centers

投资收益率(ROI)。
剩余收益(RI)。

Managers of investment centers have responsibility for costs, revenues and capital investment. Divisional performance is commonly measured using the following:

a) Return on investment (ROI).
b) Residual income (RI).

$$ROI = \frac{\text{controllable profit}}{\text{controllable investment}} \times 100\%$$

RI = Controllable profit − Interest charge on controllable

investment

ROI shows how much profit has been made in relation to the amount of capital invested.

剩餘收益是一個絕對值，是企業利潤減去投資額按規定(或預期)的最低收益率計算的投資收益後剩下的那部分。

RI is a measure of the centers' profits after deducting a notional or imputed interest cost, say, the income generated by a firm after accounting for the true cost of capital.「Residual」means in excess of any opportunity costs measured relative to the book value of Shareholders' equity.

Despite the formulas shown above, ROI and RI can be calculated in different ways.

To illustrate, let's suppose that KLS company has two investment centers, electronics division and medical supplies division, which show results for the year in table 10.3 as follow (the company's cost of capital is 10%):

Table 10.3　　　　　　　　　**Investment Results**

	Electronics division	Medical supplies division
Controllable profit($)	60,000	30,000
Capital employed($)	400,000	120,000
Cost of capital	10%	10%
ROI	15%	25%
RI ($)	20,000	18,000

Advantages of ROI and RI:
ROI can be used for comparing the returns of dissimilar business; Use of ROI encourages manager focus on cost efficiency and operating asset efficiency;
RI is an absolute measure, which overcome some of the dysfunctional consequences of ROI (relative measure). RI can be related to the NPV of a project and supports the NPV approach, therefore RI is more likely to encourage goal congruence comparing to ROI.

Disadvantages of ROI and RI:
Although ROI is the most widely used financial measure of divisional performance, it does present problems, so does RI. ROI can lead to lack of goal congruence, and encourages managers to focus on the short run at the expense of the long run. Assuming in the KLS company example, a new

proposal which requires $100,000 investment is expected to offer return of 20%, how would the managers from electronics and medical supplies divisions respond respectively?

Both ROI and RI involve a cost of capital figure which is difficult to estimate.

10.4 Economic Value Added

10.4.1 Concept of EVA

經濟增加值(EVA):稅後淨營業利潤減去全部資本成本後的剩餘收入。其核心是企業的盈利只有高於其資本成本(包括股權成本和債務成本)時才會為股東創造價值。

EVA, also referred to as economic profit, is a measure of a company's financial performance based on the residual income calculated by deducting cost of capital from its operating profit.

The measure was devised by Stern Stewart & Co. in early 1990's[1]. It has quickly adopted by a number of companies for performance evaluation. For example, Coca-Cola had EVA of $1,562 million in 1999, General Electric and Microsoft experienced significant increases in wealth for 1999 according to EVA while other companies, such as Pepsico and HP showed decreased in wealth[2].

The underlying idea is that investors require a rate of return from their resources — i.e. equity — under the control of the firm's management, compensating them for their opportunity cost and accounting for the level of risk resulting. This rate of return is the cost of equity, and a formal equity cost must be subtracted from net income. Consequently, to create shareholder value, management must generate returns at least as great as this cost. Thus, although a company may report a profit on its income statement, it may actually be economically unprofitable from shareholders perspective.

The management creates value to shareholders only if the net profits generated excess the cost of equity.

The basic formula for EVA is as follows:

---

[1] EVA is a registered trademark of Stern Stewart and Company.
[2] Geoffrey Colvin. America Best & Worst Wealth Creators[J]. Fortune, 2000(12):207-216.

10.4.2 Calculation of EVA

經過調整的稅後淨營業利潤（NOPAT），調整項目根據業務不同可以非常多項，但常見的調整項目有商譽的攤銷、品牌建設、員工培訓等。

EVA = Net Operating Profit After Taxes (NOPAT) − (Capital × Cost of Capital)

NOPAT is the net operating profit after tax, with adjustments and translations. Capital is the amount of cash invested in the business, net of depreciation, or it can be calculated as the sum of interest-bearing debt and equity also with adjustments. Stern Stewart have stated that there are potentially over 160 adjustments that could be made but in practice only five or seven key ones are made, depending on the company and the industry it competes in. It is generally adjusted for the amortization of goodwill, the capitalization of brand advertising and other non-cash items. For example, R&D, employee training, and so on, should be included in capital because they meant to have a long-term pay-off despite the fact that they are classified as accounting expenses.

股權資金的成本是隱形成本，不反應在會計報表上。

The cost of capital is the minimum rate of return on capital required by both shareholders and debt holders. The former that is invisible on financial reports is often deliberately ignored by management.

Example 10.1

Calculate Hamma's EVA for 20×4 based on the following information:

Income statements of Hamma for 20×4

	$
Revenue	400
profit before tax	96
income tax	(29)
profit after tax	67
dividends	(23)
retained earnings	44

Statements of financial position of Hamma for 20×4

	$		$
Non-current assets	160	Long-term debt (@ pre-tax 10%)	70
Current assets	180	Equity	270
Total assets	340	Total debt and equity	340

Other information:

a) Amortization of goodwill amounted to $5m per year. The amount of goodwill written off against reserves on acquisitions in years prior to 20×4 amounted to $45.
b) The group's cost of equity was estimated to be 15%.
c) Depreciation amounted to $40 in 20×4.
d) Interest payable amounted to $6m per year in 20×4.
(e) Other non-cash expenses amounted to $12 in 20×4.
(f) The rate of taxation is 30%.
(g) Capital employed at the end of 20×3 was $279m.

Solution (table 10.4):

Table 10.4　　　　　EVA of the Hamma Group for 20×4

	$ (million)
Profit for the period (after tax)	67
Add: Goodwill amortized	5
Depreciation	40
Other non-cash expenses	12
Interest expenses (W1)	4.2
Adjusted profit	128.2
Adjusted capital employed	
Capital employed	279
Goodwill	45
Adjusted capital employed	324
Calculation of EVA	
Adjusted profit	128.2
Less: capital charge (W2)	(48.6)
EVA	79.6

Workings:

a) Interest expenses = Interest payable × (1−Tax) = 6×(1−30%) = 4.2

b) Apply cost of capital of 15% to the capital employed: 15%× 324 = 48.6

Comments: the EVA measures are positive which show increases in the real wealth of Hamma based on economic values.

10.4.3　Advantages and Disadvantages of EVA

You may notice that the EVA is another form of RI, both are residual value created by the company. The key feature of EVA is its emphasis on after-tax operating profit and the actual cost of capital. It makes managers to think more about the use of capital and the amount of capital in each business.

EVA 使企業決策與股東目標一致。

EVA is likely to encourage goal congruence. Adopting EVA can encourage managers to act in the best interest of shareholders, to make better decisions that benefit the company in long-term. For example, investment on some activities such as research and development, staff training, marketing and advertising could damage short-term profit but they are crucial for the company's future success. If EVA is used to evaluate performance and link to reward scheme, managers will have incentives to pursue these investments.

EVA 考慮了股東的資金成本。

In addition, calculation of EVA that makes cost of capital visible enables managers aware that the capital has cost and they are thus encouraged to dispose of underutilized assets that don't generate sufficient income to cover their cost of capital.

Disadvantages of EVA. EVA depends on historical data, which may be a limited guide for future.

EVA requires a large number of adjustments to be made to accounting information, and the number of adjustments required can make EVA difficult (and time consuming) to use. In addition, what adjustments should be made subject to manager's judgment, it may present the result far from the true value of the organization.

10.5　Balanced Scorecard

ROI and EVA are important measures of managerial financial performance. However, focusing on only financial figures may drive the organization pursuing short-term profits at the expenses of long-term development, other important variables such as staff training, product quality, reliability, after-sales service and customer satisfaction become vital to compete successfully in a global economic environment. To deal with this problem, Kaplan and Norton (1992) developed balance scorecard, which has been adopted extensively in business and by organization worldwide.

10.5.1 BSC Framework

平衡計分卡是戰略管理系統,從四個維度將企業的願景和戰略落實為具體計劃。

BSC is a strategic planning and performance measure system that is used to align an organization's vision and strategic objectives with is tactical business activities. It allows managers to translate the organization's vision and mission directly into meaningful financial and non-financial work plans that can be communicated to employees. BSC proposes that the organization should be viewed from the following four perspectives, with metrics developed, data collected and analyzed for each of them. Please look at figure 10.2 and table 10.5 as following:

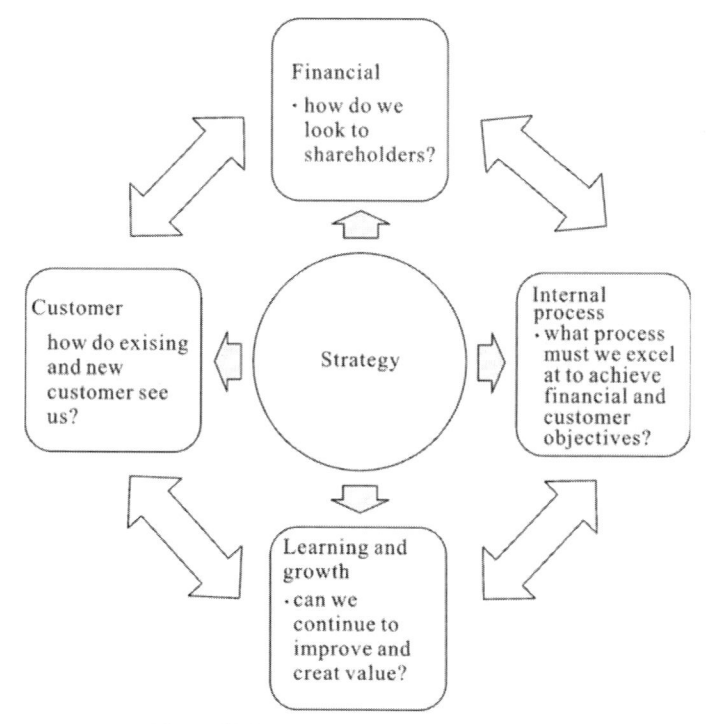

Figure 10.2 Balance Scorecard Framework

Table 10.5   Diagram of the Balance Scorecard

Perspective	Explanation
Customer	Gives rise to targets that matter to customers: cost, quality, delivery, etc
Internal Process	Aims to improve internal processes, decision making and resource utilization
Learning and growth	Considers the investment in new skills and new products that enable company maintain competitive position
Financial perspective	Traditional measures: growth, profitability and shareholder value

Each perspective includes objectives, measures of those ob-

10.5.2 Objectives, Measures and Targets

從四個方面制定目標、指標、目標值及行動方案。

jectives, target value of those measures and initiatives. In practice, companies should customize these measures to fit their own specific strategies.

The table 10.6, illustration of potential performance measures may give you an idea about how a balance scorecard may appear in different business sectors:

Table 10.6　　　　　　Potential Performance Measures

	Generic	Airline	Banking
Financial strength (Looking back)	Market share, Revenue growth, operating profit, return on equity, stock market performance	Revenues/ cost per available passenger mile, mix of full fare to discounted, average age of fleet	Outstanding loan balances, deposit balance, non-interest income
Customer service & satisfaction (Looking from the outside in)	Customer satisfaction, customer retention, quality customer service, sales from new product	Lost bag reports per 10,000 passengers denied boarding rate flight cancellation rate customer complains	Customer retention number of new customers number of products per customer Average waiting time for service
Internal operating efficiency (Looking from the inside out)	Delivery time cost process quality error rates on shipments supplier satisfaction	Load factors (% of seats occupied) utilization factors on aircraft and personnel, on-time performance	Sales calls to potential customers cross selling statistics
Learning & growth (Looking ahead)	Employee skill training availability employee satisfaction job retention amount of overtime work	Employee absenteeism worker safety statistics performance appraisals completed training hours per employee	Test results from training employee satisfaction survey

The measures selected, particularly with in the process efficiency perspective, may vary considerably with the type of organization and its objective. Once the organization has analyzed the specific and quantifiable results of the above, it may appear as table 10.7.

Table 10.7 shows a hypothetical example of a BSC for measuring the annual performance of a manufacturing firm[1].

Table 10.7　　　　　　The Balanced Scorecard

Financial		Internal process	
Sales growth	3%	Productivity	3.8%
Return on sales	6.8%	Labor turnover	13%

---

① http://www.quickmba.com/accounting/mgmt/balanced-scorecard/.

Table10. 7(continued)

Return on assets	5.1%	Ave. unit production	4 days
ROE	15.5%	Working capital/sales	10%
Gearing	64%	Capacity utilization	72%
Customers/ market		**Learning and development**	
Market share	29%	New products developed	1
No. of new customers	12,340	New markets entered	2
Product return rate	1.5%	R&D spend/ sales	2.5%
Defects	2.5%	Training spend/sales	5.6%
Order cycle time	7 days	Investment/total assets	14%

The performance of individual or responsibility center is periodically assessed by comparing actual results to the metrics in the four perspectives, then receiving feedback and adjusting objectives accordingly.

10.5.3 Characteristics of BSC

BSC translates the business strategy into four perspectives, it balances between the followings:
a) Between internal and external measures.
b) Between long-term strategy and short-term initiatives.
c) Between financial and non-financial measures.

10.5.4 Benefits of BSC

Aligning business activities with organization strategy: BSC enables organizations to clarify their vision and strategy and translate them into action.

Improving communications and performance: BSC enables the employee from bottom to top, understand how they would contribute to business strategy, and provides feedback around both internal business processes and external outcomes in order to continuously improve strategic performance and results.

After implementation in many organizations for years, the balance scorecard has evolved from performance measurement system to core strategic management system[1] that fo-

---

① Kaplan, Norton.Using the Balance Scorecard as A Strategic Management System[J]. Harvard business Review, 1996.

Kaplan, Norton.The Balance Scorecard: Translating Strategy into Action[J]. Harvard business Review, 1996.

cus on communicating and evaluating the achievement of the mission and strategy of the organization.

QUESTIONS:

1. In the last year a division's controllable return on investment was 25% and its controllable profit was $80,000. The cost of finance appropriate to the division was 18% per annum.
What was the division's controllable residual income in the last year?
A. $5,600
B. $22,400
C. $74,400
D. $76,400

2. A government body uses measures based upon the 「three Es」to the measure value for money generated by a publicly funded hospital. It considers the most important performance measure to be 「cost per successfully treated patient」.
Which of the three E's best describes the above measure?
A. Economy
B. Effectiveness
C. Efficiency
D. Externality

3. A government is looking at assessing hospitals by reference to a range of both financial and non-financial factors, one of which is survival rates for heart by-pass operation.
Which of the three E's best describes the above measure?
A. Economy
B. Effectiveness
C. Efficiency
D. Externality

4. Which of the following measures would not be appropriate for a cost center?
A. Cost per unit
B. Contribution per unit
C. Comparison of actual labour cost to budget labour cost
D. Under or over absorption of overheads

5. A government is looking at assessing state schools by reference to a range of both financial and non-financial factors, one of which is average class sizes.
Which of the three E's best describes the above measure?
A. Economy
B. Effectiveness
C. Efficiency
D. Externality

6. Copenhagen is an insurance company. Recently there has been concern that too many quotations have been sent to clients either late or containing errors. The department concerned has responded that it is understaffed, and a high proportion of current staff has recently joined the firm. The performance of this department is to be carefully monitored.
Which one of the following non-financial performance indicators would not be an appropriate measure to monitor and improve the department's performance?

A. Percentage of quotations found to contain errors when checked
B. Percentage of quotations not issued within company policy of three working days
C. Percentage of department's quota of staff actually employed
D. Percentage of budgeted number of quotations actually issued

7. For operational purposes, for a company operating a fleet of delivery vehicles, which of the following would be most useful?
A. Cost per mile run
B. Cost per driver hour
C. Cost per tonne mile
D. Cost per kg carried

8. A division has a residual income of £ 240,000 and a net profit before imputed interest of £ 640,000. If it uses a rate of 10% for computing imputed interest on its invested capital, what is its return on investment (ROI) to the nearest whole number?
A. 4%   B. 10%
C. 16%   D. 27%

9. An organisation is divided into a number of divisions, each of which operates as a profit center. Which of the following would be useful measures to monitor divisional performance?
(i) Contribution
(ii) Controllable profit
(iii) Return on investment
(iv) Residual income
A. (i) only
B. (i) and (ii) only
C. (iii) and (iv) only
D. All of them

10. HH plc monitors the % of total sales that derives from products developed in the last year. Which part of the balanced scorecard would this measure be classified under?
A. Financial perspective
B. Customer perspective
C. Internal perspective
D. Learning perspective

11. Which of the following KPIs would be used to assess the liquidity of a company?
(i) Return on capital employed
(ii) Gross profit percentage
(iii) Acid test ratio
(iv) Gearing ratio
A. (i) and (ii) only
B. (iii) only

C. (iv) only
D. (iii) and (iv) only

12. A company wants to encourage an investment centre to make new investments.
Performance measurement using which of the following KPIs would achieve this?
A. ROI   B. ROCE
C. RI   D. IRR

13. Why would a company want to encourage the use of non-financial performance indicators?
A. To encourage short termism
B. To look at the fuller picture of the business
C. To enable results to be easily manipulated to the benefit of the manager
D. To prevent goal congruence

14. K Class has calculated the following indicators
(i) Return on capital employed
(ii) Training costs as a percentage of total costs
Which of the balanced scorecard perspectives would these measures relate to?

	(i)	(ii)
A.	Financial	Financial
B.	Financial	Internal
C.	Internal	Learning and growth
D.	Financial	Learning and growth

15. Area 27 are a pizza delivery company and have asked you to suggest some performance indicators that could be used to measure the customer perspective and the internal perspective of the balanced scorecard. Which of the following would be appropriate?

	Customer	Internal
A.	Number of customer complaints	Time taken from order to delivery pizza
B.	Cost per pizza	Cost of time spent on training
C.	Number of late deliveries	Profit per pizza
D.	Cost of delivery vehicles	Gross profit percentage

Appendix: Present Value Tables and Future Value Tables

Table A Present Value of $1

Present Value

Periods	1%	2%	3%	4%	5%	6%	7%	8%	9%	10%
1	0.990,1	0.990,2	0.990,2	0.990,2	0.990,2	0.990,2	0.990,2	0.990,2	0.990,2	0.990,2
2	0.980,3	0.961,2	0.942,6	0.924,6	0.907,0	0.890,0	0.873,4	0.857,3	0.841,7	0.826,4
3	0.970,6	0.942,3	0.915,1	0.889,0	0.863,8	0.839,6	0.816,3	0.793,8	0.772,2	0.751,3
4	0.961,0	0.923,8	0.888,5	0.854,8	0.822,7	0.792,1	0.762,9	0.735,0	0.708,4	0.683,0
5	0.951,5	0.905,7	0.862,6	0.821,9	0.783,5	0.747,3	0.713,0	0.680,6	0.649,9	0.620,9
6	0.942,0	0.888,0	0.837,5	0.790,3	0.746,2	0.705,0	0.666,3	0.630,2	0.596,3	0.564,5
7	0.932,7	0.870,6	0.813,1	0.759,9	0.710,7	0.665,1	0.622,7	0.583,5	0.547,0	0.513,2
8	0.923,5	0.853,5	0.789,4	0.730,7	0.676,8	0.627,4	0.582,0	0.540,3	0.501,9	0.466,5
9	0.914,3	0.836,8	0.766,4	0.702,6	0.644,6	0.591,9	0.543,9	0.500,2	0.460,4	0.424,1
10	0.905,3	0.820,3	0.744,1	0.675,6	0.613,9	0.558,4	0.508,3	0.463,2	0.422,4	0.385,5
11	0.896,3	0.804,3	0.722,4	0.649,6	0.584,7	0.526,8	0.475,1	0.428,9	0.387,5	0.350,5
12	0.887,4	0.788,5	0.701,4	0.624,6	0.556,8	0.497,0	0.444,0	0.397,1	0.355,5	0.318,6
13	0.878,7	0.773,0	0.681,0	0.600,6	0.530,3	0.468,8	0.415,0	0.367,7	0.326,2	0.289,7
14	0.870,0	0.757,9	0.661,1	0.577,5	0.505,1	0.442,3	0.387,8	0.340,5	0.299,2	0.263,3
15	0.861,3	0.743,0	0.641,9	0.555,3	0.481,0	0.417,3	0.362,4	0.315,2	0.274,5	0.239,4
16	0.852,8	0.728,4	0.623,2	0.533,9	0.458,1	0.393,6	0.338,7	0.291,9	0.251,9	0.217,6
17	0.844,4	0.714,2	0.605,0	0.513,4	0.436,3	0.371,4	0.316,6	0.270,3	0.231,1	0.197,8
18	0.836,0	0.700,2	0.587,4	0.493,6	0.415,5	0.350,3	0.295,9	0.250,2	0.212,0	0.179,9
19	0.827,7	0.686,4	0.570,3	0.474,6	0.395,7	0.330,5	0.276,5	0.231,7	0.194,5	0.163,5
20	0.819,5	0.673,0	0.553,7	0.456,4	0.376,9	0.311,8	0.258,4	0.214,5	0.178,4	0.148,6
21	0.811,4	0.659,8	0.537,5	0.438,8	0.358,9	0.294,2	0.241,5	0.198,7	0.163,7	0.135,1
22	0.803,4	0.646,8	0.521,9	0.422,0	0.341,8	0.277,5	0.225,7	0.183,9	0.150,2	0.122,8
23	0.795,4	0.634,2	0.506,7	0.405,7	0.325,6	0.261,8	0.210,9	0.170,3	0.137,8	0.111,7
24	0.787,6	0.621,7	0.491,9	0.390,1	0.310,1	0.247,0	0.197,1	0.157,7	0.126,4	0.101,5
25	0.779,8	0.609,5	0.477,6	0.375,1	0.295,3	0.233,0	0.184,2	0.146,0	0.116,0	0.092,3
26	0.772,0	0.597,6	0.463,7	0.360,7	0.281,2	0.219,8	0.172,2	0.135,2	0.106,4	0.083,9
27	0.764,4	0.585,9	0.450,2	0.346,8	0.267,8	0.207,4	0.160,9	0.125,2	0.097,6	0.076,3
28	0.756,8	0.574,4	0.437,1	0.333,5	0.255,1	0.195,6	0.150,4	0.115,9	0.089,5	0.069,3
29	0.749,3	0.563,1	0.424,3	0.320,7	0.242,9	0.184,6	0.140,6	0.107,3	0.082,2	0.063,0
30	0.741,9	0.552,1	0.412,0	0.308,3	0.231,4	0.174,1	0.131,4	0.099,4	0.075,4	0.057,3
40	0.671,7	0.452,9	0.306,6	0.208,3	0.142,0	0.097,2	0.066,8	0.046,0	0.031,8	0.022,1
50	0.608,0	0.371,5	0.228,1	0.140,7	0.087,2	0.054,3	0.033,9	0.021,3	0.013,4	0.008,5
60	0.550,4	0.304,8	0.169,7	0.095,1	0.053,5	0.030,3	0.017,3	0.009,9	0.005,7	0.003,3

Table A Present Value of $1

Present Value

Periods	12%	14%	16%	18%	20%	25%	30%	35%	40%	45%
1	0.892,9	0.877,2	0.862,1	0.847,5	0.833,3	0.800,0	0.769,2	0.740,7	0.714,3	0.689,7
2	0.797,2	0.769,5	0.743,2	0.718,2	0.694,4	0.640,0	0.591,7	0.548,7	0.510,2	0.475,6
3	0.711,8	0.675,0	0.640,7	0.608,6	0.578,7	0.512,0	0.455,2	0.406,4	0.364,4	0.328,0
4	0.635,5	0.592,1	0.552,3	0.515,8	0.482,3	0.409,6	0.350,1	0.301,1	0.260,3	0.226,2
5	0.567,4	0.519,4	0.476,1	0.437,1	0.401,9	0.327,7	0.269,3	0.223,0	0.185,9	0.156,0
6	0.506,6	0.455,6	0.410,4	0.370,4	0.334,9	0.262,1	0.207,2	0.165,2	0.132,8	0.107,6
7	0.452,3	0.399,6	0.353,8	0.313,9	0.279,1	0.209,7	0.159,4	0.122,4	0.094,9	0.074,2
8	0.403,9	0.350,6	0.305,0	0.266,0	0.232,6	0.167,8	0.122,6	0.090,6	0.067,8	0.051,2
9	0.360,6	0.307,5	0.263,0	0.225,5	0.193,8	0.134,2	0.094,3	0.067,1	0.048,4	0.035,3
10	0.322,0	0.269,7	0.226,7	0.191,1	0.161,5	0.107,4	0.072,5	0.049,7	0.034,6	0.024,3
11	0.287,5	0.236,6	0.195,4	0.161,9	0.134,6	0.085,9	0.055,8	0.036,8	0.024,7	0.016,8
12	0.256,7	0.207,6	0.168,5	0.137,2	0.112,2	0.068,7	0.042,9	0.027,3	0.017,6	0.011,6
13	0.229,2	0.182,1	0.145,2	0.116,3	0.093,5	0.055,0	0.033,0	0.020,2	0.012,6	0.008,0
14	0.204,6	0.159,7	0.125,2	0.098,5	0.077,9	0.044,0	0.025,4	0.015,0	0.009,0	0.005,5
15	0.182,7	0.140,1	0.107,9	0.083,5	0.064,9	0.035,2	0.019,5	0.011,1	0.006,4	0.003,8
16	0.163,1	0.122,9	0.093,0	0.070,8	0.054,1	0.028,1	0.015,0	0.008,2	0.004,6	0.002,6
17	0.145,6	0.107,8	0.080,2	0.060,0	0.045,1	0.022,5	0.011,6	0.006,1	0.003,3	0.001,8
18	0.130,0	0.094,6	0.069,1	0.050,8	0.037,6	0.018,0	0.008,9	0.004,5	0.002,3	0.001,2
19	0.116,1	0.082,9	0.059,6	0.043,1	0.031,3	0.014,4	0.006,8	0.003,3	0.001,7	0.000,9
20	0.103,7	0.072,8	0.051,4	0.036,5	0.026,1	0.011,5	0.005,3	0.002,5	0.001,2	0.000,6
21	0.092,6	0.063,8	0.044,3	0.030,9	0.021,7	0.009,2	0.004,0	0.001,8	0.000,9	0.000,4
22	0.082,6	0.056,0	0.038,2	0.026,2	0.018,1	0.007,4	0.003,1	0.001,4	0.000,6	0.000,3
23	0.073,8	0.049,1	0.032,9	0.022,2	0.015,1	0.005,9	0.002,4	0.001,0	0.000,4	0.000,2
24	0.065,9	0.043,1	0.028,4	0.018,8	0.012,6	0.004,7	0.001,8	0.000,7	0.000,3	0.000,1
25	0.058,8	0.037,8	0.024,5	0.016,0	0.010,5	0.003,8	0.001,4	0.000,6	0.000,2	0.000,1
26	0.052,5	0.033,1	0.021,1	0.013,5	0.008,7	0.003,0	0.001,1	0.000,4	0.000,2	0.000,1
27	0.046,9	0.029,1	0.018,2	0.011,5	0.007,3	0.002,4	0.000,8	0.000,3	0.000,1	0.000,0
28	0.041,9	0.025,5	0.015,7	0.009,7	0.006,1	0.001,9	0.000,6	0.000,2	0.000,1	0.000,0
29	0.037,4	0.022,4	0.013,5	0.008,2	0.005,1	0.001,5	0.000,5	0.000,2	0.000,1	0.000,0
30	0.033,4	0.019,6	0.011,6	0.007,0	0.004,2	0.001,2	0.000,4	0.000,1	0.000,0	0.000,0
40	0.010,7	0.005,3	0.002,6	0.001,3	0.000,7	0.000,1	0.000,0	0.000,0	0.000,0	0.000,0
50	0.003,5	0.001,4	0.000,6	0.000,3	0.000,1	0.000,0	0.000,0	0.000,0	0.000,0	0.000,0
60	0.001,1	0.000,4	0.000,1	0.000,0	0.000,0	0.000,0	0.000,0	0.000,0	0.000,0	0.000,0

Table B Future Value of $1

Future Value

Periods	1%	2%	3%	4%	5%	6%	7%	8%	9%	10%
1	1.010,0	1.020,0	1.030,0	1.040,0	1.050,0	1.060,0	1.070,0	1.080,0	1.090,0	1.100,0
2	1.020,1	1.040,4	1.060,9	1.081,6	1.102,5	1.123,6	1.144,9	1.166,4	1.188,1	1.210,0
3	1.030,3	1.061,2	1.092,7	1.124,9	1.157,6	1.191,0	1.225,0	1.259,7	1.295,0	1.331,0
4	1.040,6	1.082,4	1.125,5	1.169,9	1.215,5	1.262,5	1.310,8	1.360,5	1.411,6	1.464,1
5	1.051,0	1.104,1	1.159,3	1.216,7	1.276,3	1.338,2	1.402,6	1.469,3	1.538,6	1.610,5
6	1.061,5	1.126,2	1.194,1	1.265,3	1.340,1	1.418,5	1.500,7	1.586,9	1.677,1	1.771,6
7	1.072,1	1.148,7	1.229,9	1.315,9	1.407,1	1.503,6	1.605,8	1.713,8	1.828,0	1.948,7
8	1.082,9	1.171,7	1.266,8	1.368,6	1.477,5	1.593,8	1.718,2	1.850,9	1.992,6	2.143,6
9	1.093,7	1.195,1	1.304,8	1.423,3	1.551,3	1.689,5	1.838,5	1.999,0	2.171,9	2.357,9
10	1.104,6	1.219,0	1.343,9	1.480,2	1.628,9	1.790,8	1.967,2	2.158,9	2.367,4	2.593,7
11	1.115,7	1.243,4	1.384,2	1.539,5	1.710,3	1.898,3	2.104,9	2.331,6	2.580,4	2.853,1
12	1.126,8	1.268,2	1.425,8	1.601,0	1.795,9	2.012,2	2.252,2	2.518,2	2.812,7	3.138,4
13	1.138,1	1.293,6	1.468,5	1.665,1	1.885,6	2.132,9	2.409,8	2.719,6	3.065,8	3.452,3
14	1.149,5	1.319,5	1.512,6	1.731,7	1.979,9	2.260,9	2.578,5	2.937,2	3.341,7	3.797,5
15	1.161,0	1.345,9	1.558,0	1.800,9	2.078,9	2.396,6	2.759,0	3.172,2	3.642,5	4.177,2
16	1.172,6	1.372,8	1.604,7	1.873,0	2.182,9	2.540,4	2.952,2	3.425,9	3.970,3	4.595,0
17	1.184,3	1.400,2	1.652,8	1.947,9	2.292,0	2.692,8	3.158,8	3.700,0	4.327,6	5.054,5
18	1.196,1	1.428,2	1.702,4	2.025,8	2.406,6	2.854,3	3.379,9	3.996,0	4.717,1	5.559,9
19	1.208,1	1.456,8	1.753,5	2.106,8	2.527,0	3.025,6	3.616,5	4.315,7	5.141,7	6.115,9
20	1.220,2	1.485,9	1.806,1	2.191,1	2.653,3	3.207,1	3.869,7	4.661,0	5.604,4	6.727,5
21	1.232,4	1.515,7	1.860,3	2.278,8	2.786,0	3.399,6	4.140,6	5.033,8	6.108,8	7.400,2
22	1.244,7	1.546,0	1.916,1	2.369,9	2.925,3	3.603,5	4.430,4	5.436,5	6.658,6	8.140,3
23	1.257,2	1.576,9	1.973,6	2.464,7	3.071,5	3.819,7	4.740,5	5.871,5	7.257,9	8.954,3
24	1.269,7	1.608,4	2.032,8	2.563,3	3.225,1	4.048,9	5.072,4	6.341,2	7.911,1	9.849,7
25	1.282,4	1.640,6	2.093,8	2.665,8	3.386,4	4.291,9	5.427,4	6.848,5	8.623,1	10.835
26	1.295,3	1.673,4	2.156,6	2.772,5	3.555,7	4.549,4	5.807,4	7.396,4	9.399,2	11.918
27	1.308,2	1.706,9	2.221,3	2.883,4	3.733,5	4.822,3	6.213,9	7.988,1	10.245	13.110
28	1.321,3	1.741,0	2.287,9	2.998,7	3.920,1	5.111,7	6.648,8	8.627,1	11.167	14.421
29	1.334,5	1.775,8	2.356,6	3.118,7	4.116,1	5.418,4	7.114,3	9.317,3	12.172	15.863
30	1.347,8	1.811,4	2.427,3	3.243,4	4.321,9	5.743,5	7.612,3	10.063	13.268	17.449
40	1.488,9	2.208,0	3.262,0	4.801,0	7.040,0	10.286	14.974	21.725	31.409	45.259
50	1.644,6	2.691,6	4.383,9	7.106,7	11.467	18.420	29.457	46.902	74.358	117.39
60	1.816,7	3.281,0	5.891,6	10.520	18.679	32.988	57.946	101.26	176.03	304.48

Table B Future Value of $1

Future Value

Periods	12%	14%	16%	18%	20%	25%	30%	35%	40%	45%
1	1.120,0	1.140,0	1.160,0	1.180,0	1.200,0	1.250,0	1.300,0	1.350,0	1.400,0	1.450,0
2	1.254,4	1.299,6	1.345,6	1.392,4	1.440,0	1.562,5	1.690,0	1.822,5	1.960,0	2.102,5
3	1.404,9	1.481,5	1.560,9	1.643,0	1.728,0	1.953,1	2.197,0	2.460,4	2.744,0	3.048,6
4	1.573,5	1.689,0	1.810,6	1.938,8	2.073,6	2.441,4	2.856,1	3.321,5	3.841,6	4.420,5
5	1.762,3	1.925,4	2.100,3	2.287,8	2.488,3	3.051,8	3.712,9	4.484,0	5.378,2	6.409,7
6	1.973,8	2.195,0	2.436,4	2.699,6	2.986,0	3.814,7	4.826,8	6.053,4	7.529,5	9.294,1
7	2.210,7	2.502,3	2.826,2	3.185,5	3.583,2	4.768,4	6.274,9	8.172,2	10.541,4	13.476,5
8	2.476,0	2.852,6	3.278,4	3.758,9	4.299,8	5.960,5	8.157,3	11.032,4	14.757,9	19.540,9
9	2.773,1	3.251,9	3.803,0	4.435,5	5.159,8	7.450,6	10.604,5	14.893,7	20.661,0	28.334,3
10	3.105,8	3.707,2	4.411,4	5.233,8	6.191,7	9.313,2	13.785,8	20.106,6	28.925,5	41.084,7
11	3.478,5	4.226,2	5.117,3	6.175,9	7.430,1	11.642	17.922	27.144	40.496	59.573
12	3.896,0	4.817,9	5.936,0	7.287,6	8.916,1	14.552	23.298	36.644	56.694	86.381
13	4.363,5	5.492,4	6.885,8	8.599,4	10.699	18.190	30.288	49.470	79.371	125.25
14	4.887,1	6.261,3	7.987,5	10.147	12.839	22.737	39.374	66.784	111.12	181.62
15	5.473,6	7.137,9	9.265,5	11.974	15.407	28.422	51.186	90.158	155.57	263.34
16	6.130,4	8.137,2	10.748	14.129	18.488	35.527	66.542	121.71	217.80	381.85
17	6.866,0	9.276,5	12.468	16.672	22.186	44.409	86.504	164.31	304.91	553.68
18	7.690,0	10.575	14.463	19.673	26.623	55.511	112.46	221.82	426.88	802.83
19	8.612,8	12.056	16.777	23.214	31.948	69.389	146.19	299.46	597.63	1,164.1
20	9.646,3	13.743	19.461	27.393	38.338	86.736	190.05	404.27	836.68	1,688.0
21	10.804	15.668	22.574	32.324	46.005	108.42	247.06	545.77	1,171.4	2,447.5
22	12.100	17.861	26.186	38.142	55.206	135.53	321.18	736.79	1,639.9	3,548.9
23	13.552	20.362	30.376	45.008	66.247	169.41	417.54	994.66	2,295.9	5,145.9
24	15.179	23.212	35.236	53.109	79.497	211.76	542.80	1,342.8	3,214.2	7,461.6
25	17.000	26.462	40.874	62.669	95.396	264.70	705.64	1,812.8	4,499.9	10,819.3
26	19.040	30.167	47.414	73.949	114.48	330.87	917.33	2,447.2	6,299.8	15,688.0
27	21.325	34.390	55.000	87.260	137.37	413.59	1,192.5	3,303.8	8,819.8	22,747.6
28	23.884	39.204	63.800	102.97	164.84	516.99	1,550.3	4,460.1	12,347.7	32,984.1
29	26.750	44.693	74.009	121.50	197.81	646.23	2,015.4	6,021.1	17,286.7	47,826.9
30	29.960	50.950	85.850	143.37	237.38	807.79	2,620.0	8,128.5	24,201.4	69,349.0
40	93.051	188.88	378.72	750.38	1,469.8	7,523	36,119	163,437	700,038	2,849,181
50	289.00	700.23	1,670.7	3,927	9,100.4	70,065	497,929	3,286,158	20,248,916	117,057,734
60	897.60	2,595.9	7,370.2	20,555	56,348	652,530	6,864,377	66,073,317	585,709,328	4,809,280,790

Table C Present Value of Annuity of $1

Present Value

Periods	1%	2%	3%	4%	5%	6%	7%	8%	9%	10%
1	0.990,1	0.980,4	0.970,9	0.961,5	0.952,4	0.943,4	0.934,6	0.925,9	0.917,4	0.909,1
2	1.970,4	1.941,6	1.913,5	1.886,1	1.859,4	1.833,4	1.808,0	1.783,3	1.759,1	1.735,5
3	2.941,0	2.883,9	2.828,6	2.775,1	2.723,2	2.673,0	2.624,3	2.577,1	2.531,3	2.486,9
4	3.902,0	3.807,7	3.717,1	3.629,9	3.546,0	3.465,1	3.387,2	3.312,1	3.239,7	3.169,9
5	4.853,4	4.713,5	4.579,7	4.451,8	4.329,5	4.212,4	4.100,2	3.992,7	3.889,7	3.790,8
6	5.795,5	5.601,4	5.417,2	5.242,1	5.075,7	4.917,3	4.766,5	4.622,9	4.485,9	4.355,3
7	6.728,2	6.472,0	6.230,3	6.002,1	5.786,4	5.582,4	5.389,3	5.206,4	5.033,0	4.868,4
8	7.651,7	7.325,5	7.019,7	6.732,7	6.463,2	6.209,8	5.971,3	5.746,6	5.534,8	5.334,9
9	8.566,0	8.162,2	7.786,1	7.435,3	7.107,8	6.801,7	6.515,2	6.246,9	5.995,2	5.759,0
10	9.471,3	8.982,6	8.530,2	8.110,9	7.721,7	7.360,1	7.023,6	6.710,1	6.417,7	6.144,6
11	10.367,6	9.786,8	9.252,6	8.760,5	8.306,4	7.886,9	7.498,7	7.139,0	6.805,2	6.495,1
12	11.255,1	10.575,3	9.954,0	9.385,1	8.863,3	8.383,8	7.942,7	7.536,1	7.160,7	6.813,7
13	12.133,7	11.348,4	10.635,0	9.985,6	9.393,6	8.852,7	8.357,7	7.903,8	7.486,9	7.103,4
14	13.003,7	12.106,2	11.296,1	10.563,1	9.898,6	9.295,0	8.745,5	8.244,2	7.786,2	7.366,7
15	13.865,1	12.849,3	11.937,9	11.118,4	10.379,7	9.712,2	9.107,9	8.559,5	8.060,7	7.606,1
16	14.717,9	13.577,7	12.561,1	11.652,3	10.837,8	10.105,9	9.446,6	8.851,4	8.312,6	7.823,7
17	15.562,3	14.291,9	13.166,1	12.165,7	11.274,1	10.477,3	9.763,2	9.121,6	8.543,6	8.021,6
18	16.398,3	14.992,0	13.753,5	12.659,3	11.689,6	10.827,6	10.059,1	9.371,9	8.755,6	8.201,4
19	17.226,0	15.678,5	14.323,8	13.133,9	12.085,3	11.158,1	10.335,6	9.603,6	8.950,1	8.364,9
20	18.045,6	16.351,4	14.877,5	13.590,3	12.462,2	11.469,9	10.594,0	9.818,1	9.128,5	8.513,6
21	18.857,0	17.011,2	15.415,0	14.029,2	12.821,2	11.764,1	10.835,5	10.016,8	9.292,2	8.648,7
22	19.660,4	17.658,0	15.936,9	14.451,1	13.163,0	12.041,6	11.061,2	10.200,7	9.442,4	8.771,5
23	20.455,8	18.292,2	16.443,6	14.856,8	13.488,6	12.303,4	11.272,2	10.371,1	9.580,2	8.883,2
24	21.243,4	18.913,9	16.935,5	15.247,0	13.798,6	12.550,4	11.469,3	10.528,8	9.706,6	8.984,7
25	22.023,2	19.523,5	17.413,1	15.622,1	14.093,9	12.783,4	11.653,6	10.674,8	9.822,6	9.077,0
26	22.795,2	20.121,0	17.876,8	15.982,8	14.375,2	13.003,2	11.825,8	10.810,0	9.929,0	9.160,9
27	23.559,6	20.706,9	18.327,0	16.329,6	14.643,0	13.210,5	11.986,7	10.935,2	10.026,6	9.237,2
28	24.316,4	21.281,3	18.764,1	16.663,1	14.898,1	13.406,2	12.137,1	11.051,1	10.116,1	9.306,6
29	25.065,8	21.844,4	19.188,5	16.983,7	15.141,1	13.590,7	12.277,7	11.158,4	10.198,3	9.369,6
30	25.807,7	22.396,5	19.600,4	17.292,0	15.372,5	13.764,8	12.409,0	11.257,8	10.273,7	9.426,9
40	32.834,7	27.355,5	23.114,8	19.792,8	17.159,1	15.046,3	13.331,7	11.924,6	10.757,4	9.779,1
50	39.196,1	31.423,6	25.729,8	21.482,2	18.255,9	15.761,9	13.800,7	12.233,5	10.961,7	9.914,8
60	44.955,0	34.760,9	27.675,6	22.623,5	18.929,3	16.161,4	14.039,2	12.376,6	11.048,0	9.967,2

Table C Present Value of Annuity of $1

Present Value

Periods	12%	14%	16%	18%	20%	25%	30%	35%	40%	45%
1	0.892,9	0.877,2	0.862,1	0.847,5	0.833,3	0.800,0	0.769,2	0.740,7	0.714,3	0.689,7
2	1.690,1	1.646,7	1.605,2	1.565,6	1.527,8	1.440,0	1.360,9	1.289,4	1.224,5	1.165,3
3	2.401,8	2.321,6	2.245,9	2.174,3	2.106,5	1.952,0	1.816,1	1.695,9	1.588,9	1.493,3
4	3.037,3	2.913,7	2.798,2	2.690,1	2.588,7	2.361,6	2.166,2	1.996,9	1.849,2	1.719,5
5	3.604,8	3.433,1	3.274,3	3.127,2	2.990,6	2.689,3	2.435,6	2.220,0	2.035,2	1.875,5
6	4.111,4	3.888,7	3.684,7	3.497,6	3.325,5	2.951,4	2.642,7	2.385,2	2.168,0	1.983,1
7	4.563,8	4.288,3	4.038,6	3.811,5	3.604,6	3.161,1	2.802,1	2.507,5	2.262,8	2.057,3
8	4.967,6	4.638,9	4.343,6	4.077,6	3.837,2	3.328,9	2.924,7	2.598,2	2.330,6	2.108,5
9	5.328,2	4.946,4	4.606,5	4.303,0	4.031,0	3.463,1	3.019,0	2.665,3	2.379,0	2.143,8
10	5.650,2	5.216,1	4.833,2	4.494,1	4.192,5	3.570,5	3.091,5	2.715,0	2.413,6	2.168,1
11	5.937,7	5.452,7	5.028,6	4.656,0	4.327,1	3.656,4	3.147,3	2.751,9	2.438,3	2.184,9
12	6.194,4	5.660,3	5.197,1	4.793,2	4.439,2	3.725,1	3.190,3	2.779,2	2.455,9	2.196,5
13	6.423,5	5.842,4	5.342,3	4.909,5	4.532,7	3.780,1	3.223,3	2.799,4	2.468,5	2.204,5
14	6.628,2	6.002,1	5.467,5	5.008,1	4.610,6	3.824,1	3.248,7	2.814,4	2.477,5	2.210,0
15	6.810,9	6.142,2	5.575,5	5.091,6	4.675,5	3.859,3	3.268,2	2.825,5	2.483,9	2.213,8
16	6.974,0	6.265,1	5.668,5	5.162,4	4.729,6	3.887,4	3.283,2	2.833,7	2.488,5	2.216,4
17	7.119,6	6.372,9	5.748,7	5.222,3	4.774,6	3.909,9	3.294,8	2.839,8	2.491,8	2.218,2
18	7.249,7	6.467,4	5.817,8	5.273,2	4.812,2	3.927,9	3.303,7	2.844,3	2.494,1	2.219,5
19	7.365,8	6.550,4	5.877,5	5.316,2	4.843,5	3.942,4	3.310,5	2.847,6	2.495,8	2.220,3
20	7.469,4	6.623,1	5.928,8	5.352,7	4.869,6	3.953,9	3.315,8	2.850,1	2.497,0	2.220,9
21	7.562,0	6.687,0	5.973,1	5.383,7	4.891,3	3.963,1	3.319,8	2.851,9	2.497,9	2.221,3
22	7.644,6	6.742,9	6.011,3	5.409,9	4.909,4	3.970,5	3.323,0	2.853,3	2.498,5	2.221,6
23	7.718,4	6.792,1	6.044,2	5.432,1	4.924,5	3.976,4	3.325,4	2.854,3	2.498,9	2.221,8
24	7.784,3	6.835,1	6.072,6	5.450,9	4.937,1	3.981,1	3.327,2	2.855,0	2.499,2	2.221,9
25	7.843,1	6.872,9	6.097,1	5.466,9	4.947,6	3.984,9	3.328,6	2.855,6	2.499,4	2.222,0
26	7.895,7	6.906,1	6.118,2	5.480,4	4.956,3	3.987,9	3.329,7	2.856,0	2.499,6	2.222,1
27	7.942,6	6.935,2	6.136,4	5.491,9	4.963,6	3.990,3	3.330,5	2.856,3	2.499,7	2.222,1
28	7.984,4	6.960,7	6.152,0	5.501,6	4.969,7	3.992,3	3.331,2	2.856,5	2.499,8	2.222,2
29	8.021,8	6.983,0	6.165,6	5.509,8	4.974,7	3.993,8	3.331,7	2.856,7	2.499,9	2.222,2
30	8.055,2	7.002,7	6.177,2	5.516,8	4.978,9	3.995,0	3.332,1	2.856,8	2.499,9	2.222,2
40	8.243,8	7.105,0	6.233,5	5.548,2	4.996,6	3.999,5	3.333,2	2.857,1	2.500,0	2.222,2
50	8.304,5	7.132,7	6.246,3	5.554,1	4.999,5	3.999,9	3.333,3	2.857,1	2.500,0	2.222,2
60	8.324,0	7.140,1	6.249,2	5.555,3	4.999,9	4.000,0	3.333,3	2.857,1	2.500,0	2.222,2

Table D Future Value of Annuity of $1

Future Value

Periods	1%	2%	3%	4%	5%	6%	7%	8%	9%	10%
1	1.000,0	1.000,0	1.000,0	1.000,0	1.000,0	1.000,0	1.000,0	1.000,0	1.000,0	1.000,0
2	2.010,0	2.020,0	2.030,0	2.040,0	2.050,0	2.060,0	2.070,0	2.080,0	2.090,0	2.100,0
3	3.030,1	3.060,4	3.090,9	3.121,6	3.152,5	3.183,6	3.214,9	3.246,4	3.278,1	3.310,0
4	4.060,4	4.121,6	4.183,6	4.246,5	4.310,1	4.374,6	4.439,9	4.506,1	4.573,1	4.641,0
5	5.101,0	5.204,0	5.309,1	5.416,3	5.525,6	5.637,1	5.750,7	5.866,6	5.984,7	6.105,1
6	6.152,0	6.308,1	6.468,4	6.633,0	6.801,9	6.975,3	7.153,3	7.335,9	7.523,3	7.715,6
7	7.213,5	7.434,3	7.662,5	7.898,3	8.142,0	8.393,8	8.654,0	8.922,8	9.200,4	9.487,2
8	8.285,7	8.583,0	8.892,3	9.214,2	9.549,1	9.897,5	10.259,8	10.636,6	11.028,5	11.435,9
9	9.368,5	9.754,6	10.159,1	10.582,8	11.026,6	11.491,3	11.978,0	12.487,6	13.021,0	13.579,5
10	10.462,2	10.949,7	11.463,9	12.006,1	12.577,9	13.180,8	13.816,4	14.486,6	15.192,9	15.937,4
11	11.566,8	12.168,7	12.807,8	13.486,4	14.206,8	14.971,6	15.783,6	16.645,5	17.560,3	18.531,2
12	12.682,5	13.412,1	14.192,0	15.025,8	15.917,1	16.869,9	17.888,5	18.977,1	20.140,7	21.384,3
13	13.809,3	14.680,3	15.617,8	16.626,8	17.713,0	18.882,1	20.140,6	21.495,3	22.953,4	24.522,7
14	14.947,4	15.973,9	17.086,3	18.291,9	19.598,6	21.015,1	22.550,5	24.214,9	26.019,2	27.975,0
15	16.096,9	17.293,4	18.598,9	20.023,6	21.578,6	23.276,0	25.129,0	27.152,1	29.360,9	31.772,5
16	17.257,9	18.639,3	20.156,9	21.824,5	23.657,5	25.672,5	27.888,1	30.324,3	33.003,4	35.949,7
17	18.430,4	20.012,1	21.761,6	23.697,5	25.840,4	28.212,9	30.840,2	33.750,2	36.973,7	40.544,7
18	19.614,7	21.412,3	23.414,4	25.645,4	28.132,4	30.905,7	33.999,0	37.450,2	41.301,3	45.599,2
19	20.810,9	22.840,6	25.116,9	27.671,2	30.539,0	33.760,0	37.379,0	41.446,3	46.018,5	51.159,1
20	22.019,0	24.297,4	26.870,4	29.778,1	33.066,0	36.785,6	40.995,5	45.762,0	51.160,1	57.275,0
21	23.239,2	25.783,3	28.676,5	31.969,2	35.719,3	39.992,7	44.865,2	50.422,9	56.764,5	64.002,5
22	24.471,6	27.299,0	30.536,8	34.248,0	38.505,2	43.392,3	49.005,7	55.456,8	62.873,3	71.402,7
23	25.716,3	28.845,0	32.452,9	36.617,9	41.430,5	46.995,8	53.436,1	60.893,3	69.531,9	79.543,0
24	26.973,5	30.421,9	34.426,5	39.082,6	44.502,0	50.815,6	58.176,7	66.764,8	76.789,8	88.497,3
25	28.243,2	32.030,3	36.459,3	41.645,9	47.727,1	54.864,5	63.249,0	73.105,9	84.700,9	98.347,1
26	29.525,6	33.67	38.55	44.31	51.11	59.16	68.68	79.95	93.32	109.18
27	30.820,9	35.34	40.71	47.08	54.67	63.71	74.48	87.35	102.72	121.10
28	32.129,1	37.05	42.93	49.97	58.40	68.53	80.70	95.34	112.97	134.21
29	33.450,4	38.79	45.22	52.97	62.32	73.64	87.35	103.97	124.14	148.63
30	34.784,9	40.57	47.58	56.08	66.44	79.06	94.46	113.28	136.31	164.49
40	48.886,4	60.40	75.40	95.03	120.80	154.76	199.64	259.06	337.88	442.59
50	64.463,2	84.58	112.80	152.67	209.35	290.34	406.53	573.77	815.08	1,163.91
60	81.669,7	114.05	163.05	237.99	353.58	533.13	813.52	1,253.21	1,944.79	3,034.82

Table D Future Value of Annuity of $1

Future Value

Periods	12%	14%	16%	18%	20%	25%	30%	35%	40%	45%
1	1,000.0	1,000.0	1,000.0	1,000.0	1,000.0	1,000.0	1,000.0	1,000.0	1,000.0	1,000.0
2	2,120.0	2,140.0	2,160.0	2,180.0	2,200.0	2,250.0	2,300.0	2,350.0	2,400.0	2,450.0
3	3,374.4	3,439.6	3,505.6	3,572.4	3,640.0	3,812.5	3,990.0	4,172.5	4,360.0	4,552.5
4	4,779.3	4,921.1	5,066.5	5,215.4	5,368.0	5,765.6	6,187.0	6,632.9	7,104.0	7,601.1
5	6,352.8	6,610.1	6,877.1	7,154.2	7,441.6	8,207.0	9,043.1	9,954.4	10,945.6	12,021.6
6	8,115.2	8,535.5	8,977.5	9,442.0	9,929.9	11,258.8	12,756.0	14,438.4	16,323.8	18,431.4
7	10,089.0	10,730.5	11,413.9	12,141.5	12,915.9	15,073.5	17,582.8	20,491.9	23,853.4	27,725.5
8	12,299.7	13,232.8	14,240.1	15,327.0	16,499.1	19,841.9	23,857.7	28,664.0	34,394.7	41,201.9
9	14,775.7	16,085.3	17,518.5	19,085.9	20,798.9	25,802.3	32,015.0	39,696.4	49,152.6	60,742.8
10	17,548.7	19,337.3	21,321.5	23,521.3	25,958.7	33,252.9	42,619.5	54,590.2	69,813.7	89,077.1
11	20,654.6	23,044.5	25,732.9	28,755.1	32,150.4	42,566.1	56,405.3	74,696.7	98,739.1	130,161.8
12	24,133.1	27,270.7	30,850.2	34,931.1	39,580.5	54,207.7	74,327.0	101,840.6	139,234.8	189,734.6
13	28,029.1	32,088.7	36,786.2	42,218.7	48,496.6	68,759.6	97,625.0	138,484.8	195,928.7	276,115.1
14	32,392.6	37,581.1	43,672.0	50,818.0	59,195.9	86,949.5	127,912.5	187,954.4	275,300.2	401,367.0
15	37,279.7	43,842.4	51,659.5	60,965.3	72,035.1	109,686.8	167,286.3	254,738.5	386,420.2	582,982.1
16	42,753.3	50,980.4	60,925.0	72,939.0	87,442.1	138,108.5	218,472.2	344.90	541.99	846.32
17	48,883.7	59,117.6	71,673.0	87,068.0	105,930.6	173,635.7	285,013.9	466.61	759.78	1,228.17
18	55,749.7	68,394.1	84,140.7	103,740.3	128,116.7	218,044.6	371,518.0	630.92	1,064.70	1,781.85
19	63,439.7	78,969.2	98,603.2	123,413.5	154,740.0	273,555.8	483,973.4	852.75	1,491.58	2,584.68
20	72,052.4	91,024.9	115,379.7	146,628.0	186,688.0	342,944.7	630,165.5	1,152.21	2,089.21	3,748.78
21	81.70	104.77	134.84	174.02	225.03	429.68	820.22	1,556.48	2,925.89	5,436.73
22	92.50	120.44	157.41	206.34	271.03	538.10	1,067.28	2,102.25	4,097.24	7,884.26
23	104.60	138.30	183.60	244.49	326.24	673.63	1,388.46	2,839.04	5,737.14	11,433
24	118.16	158.66	213.98	289.49	392.48	843.03	1,806.00	3,833.71	8,033.00	16,579
25	133.33	181.87	249.21	342.60	471.98	1,054.79	2,348.80	5,176.50	11,247.20	24,041
26	150.33	208.33	290.09	405.27	567.38	1,319.49	3,054.44	6,989.28	15,747.08	34,860
27	169.37	238.50	337.50	479.22	681.85	1,650.36	3,971.78	9,436.53	22,046.91	50,548
28	190.70	272.89	392.50	566.48	819.22	2,063.95	5,164.31	12,740.31	30,866.67	73,296
29	214.58	312.09	456.30	669.45	984.07	2,580.94	6,714.60	17,200.42	43,214.34	106,280
30	241.33	356.79	530.31	790.95	1,181.88	3,227.17	8,729.99	23,221.57	60,501.08	154,107
40	767.09	1,342.03	2,360.76	4,163.21	7,343.86	30,088.66	120,392.88	466,960.38	1,750,091.74	6,331,512
50	2,400.02	4,994.52	10,436	21,813	45,497	280,256	1,659,761	9,389,020	50,622,288	260,128,295
60	7,471.64	18,535	46,058	114,190	281,733	2,610,118	22,881,254	188,780,903	1,464,273,318	10,687,290,642

國家圖書館出版品預行編目(CIP)資料

管理會計 / 張紅云、韓衛華 主編. -- 第一版.
-- 臺北市：崧燁文化，2018.08
　面　；　公分
ISBN 978-957-681-433-4(平裝)
1.管理會計
494.74　　　　　107012254

書　名：管理會計
作　者：張紅云、韓衛華 主編
發行人：黃振庭
出版者：崧燁文化事業有限公司
發行者：崧燁文化事業有限公司
E-mail：sonbookservice@gmail.com
粉絲頁　　　　　　網　址：
地　址：台北市中正區重慶南路一段六十一號八樓815室
8F.-815, No.61, Sec. 1, Chongqing S. Rd., Zhongzheng Dist., Taipei City 100, Taiwan (R.O.C.)
電　話：(02)2370-3310　傳　真：(02) 2370-3210
總經銷：紅螞蟻圖書有限公司
地　址：台北市內湖區舊宗路二段 121 巷 19 號
電　話：02-2795-3656　傳真：02-2795-4100　網址：
印　刷：京崟彩色印刷有限公司（京峰數位）

　　本書版權為西南財經大學出版社所有授權崧博出版事業股份有限公司獨家發行電子書繁體字版。若有其他相關權利需授權請與西南財經大學出版社聯繫，經本公司授權後方得行使相關權利。

定價：400 元
發行日期：2018 年 8 月第一版
◎ 本書以POD印製發行